SpringerBriefs in Applied Sciences and Technology

T0203072

For further volumes:
http://www.springer.com/series/8884

Christine Charyton

Creative Engineering Design Assessment

Background, Directions, Manual, Scoring Guide and Uses

 Springer

Christine Charyton
Ohio State University
Columbus, OH
USA

ISSN 2191-530X   ISSN 2191-5318   (electronic)
ISBN 978-1-4471-5378-8   ISBN 978-1-4471-5379-5   (eBook)
DOI 10.1007/978-1-4471-5379-5
Springer London Heidelberg New York Dordrecht

Library of Congress Control Number: 2013942993

To all the engineers that make a difference in people's lives. The word "engineer" comes from the Latin word "Ingeniatorum" meaning "ingenious," with "gen" referring to creation, or "Genesis." The essence of the words "creativity," "create" and "engineer" stem from the act of creation.

In this light, I would also like to dedicate this book to my parents, Alex Charyton and June Charyton, for their encouragement during the book writing process. I would also like to dedicate this book to my husband, John Elliott, for encouraging me to critically process and assess my writing.

I would also like to thank the faculty at Temple University for encouraging me to pursue this path of research. I am grateful for the opportunity to promote experiential hands-on learning in my Psychology of Creativity class in the Department of Psychology at Ohio State University. The course has attracted many students from various majors across the university.

I am also grateful for the opportunity to promote experiential hands-on learning to the student initiated Seminar for Promoting Creativity and Innovation in the College of Engineering. Furthermore, I express gratitude to the Department of Neurology at the Ohio State University Wexner Medical Center for their support of my research.

Foreword

Our approach for the CEDA tool is unique. The standard accepted definition of creativity includes originality (novelty) and usefulness. However, to date, no assessments have directly measured these constructs. The CEDA is conceptualized to measure originality and usefulness specific to engineering design in addition to both divergent and convergent thinking.

Measurement is straightforward in that the assessor or judge (each rater) reads all the word descriptions then uses their own "gut instinct" to decide on the assessment of each design then looks for the word on the rubric that best describes their initial perspective. This is paired with a number. When this initial work was presented at the National Science Foundation (NSF) these aspects of measurement were well-received. This approach is unique especially in comparison with the Torrance Tests of Creative Thinking (TTCT), which is very structured and has received criticism for not reflecting "real-world" creativity.

The CEDA has been shown to be related to the Purdue Creativity Test (PCT) which was normed on engineers in industry. The PCT was developed in 1960 and has not been widely used.

Best wishes with your implementation of the CEDA.

Preface

The *Creative Engineering Design Assessment (CEDA): Background, Directions, Manual, Scoring Guide and Uses* has three chapters.

Chapter 1 contains the background and rationale for the CEDA tool. The importance of creativity in engineering design, engineering education and the core principals of engineering design are discussed.

Chapter 2 contains the CEDA manual and directions for administration and scoring. The theoretical framework, reliability and validity are also provided in this section. Components of measurement discussed include Fluency, Flexibility, Originality and Usefulness. The CEDA is available seven languages: English, Ukrainian, Spanish, Korean, French, Chinese and German. Directions for printing and scoring the tool are in this section.

Chapter 3 includes intended uses of the CEDA. The aim and overall purpose of the CEDA is to measure creativity specific to engineering design. The objective of the designs are to help enrich people's lives and to benefit humankind. The importance of creativity in STEM, Industry, NASA and the military are discussed in this section. Conclusions are also described. For further information regarding administration and scoring, please contact the author. Contact information appears at the end of the Acknowledgements section.

Best wishes with your interests and uses for the CEDA.

Acknowledgments

I would like to thank all of the following in a "we" format since so many people helped make this work possible. We are appreciative of all the people who have helped out with this research from past publications and future publications that are described in this book. Special gratitude goes to Alex Charyton and June Charyton for their continuous support and to John O. Elliott for his continuous editorial feedback during this book writing process. Thank you to William Clifton, Samantha DeDios, Stephanie Bennett, Thornton Lothrop, Mark Miller, Jean Wheasler, Laurin Turowski, Ankit Tayal, T. J. Starr, Berae McClary, and Jong-Hun Sun. We would also like to thank Jong-Hun Sun, Mike Interiano, Myroslava Mudrak, Kelly Tung-Steudler, Saebyul Lee, Magda Kolcio, and Martin Mueller for translations of the CEDA into other languages. We also thank Richard Jaga-cinski, John A. Merrill, the Snelbecker Family in memory of Glenn E. Snelbecker, and Yosef Alam for their feedback on earlier versions and revised versions of the CEDA. We thank Mohammed Rahman for statistical consultation as well as Shane Ruland, Larry Campbell, Paul Jones, and Marc Archuleta for their technical assistance and expertise regarding this project. Additional gratitude goes to Blaine Lilly and Glenn Elliott for their initial feedback and technological support, respectively, in the development of the first version of the CEDA. We also thank Judy Blau and Seth Blau for helping us find a German translator; we appreciate their support and interest in this project. My colleagues and I look forward to corporations and researchers from various universities and regions from around the world who have expressed their interest, using the CEDA.

For further information contact:

Christine Charyton, Ph.D.
Ohio State University
1835 Neil Avenue
Columbus, OH 43210
USA
charyton.1@osu.edu

Please feel free to contact Dr. Charyton if you would like additional information or training, scoring, and interpretation of the CEDA. Dr. Charyton can provide on-site training for your research, educational program, industry, NASA, or military project.

Contents

1 An Overview of the Relevance of Creative Engineering Design:
  Background 1
  1.1 The Importance of Creativity in Engineering Design 1
  1.2 Creativity in Engineering Education 2
  1.3 Core Principles of Creative Engineering Design 5
  1.4 Assessments to Measure Creativity in Engineering Design 6
  References 8

2 The CEDA Manual and Directions 11
  2.1 Introduction 11
  2.2 Recommended Resources 11
  2.3 Directions for Administering the CEDA 12
  2.4 Theoretical Framework of the CEDA 13
  2.5 Components of Scoring 14
  2.6 Properties of the CEDA 15
      2.6.1 Reliability 15
      2.6.2 Validity 15
      2.6.3 Conceptualization 17
  2.7 How to Score the CEDA 18
  2.8 Directions 18
      2.8.1 Fluency 20
      2.8.2 Flexibility 20
      2.8.3 Originality 21
      2.8.4 Usefulness 22
  2.9 Sample Design Problem 22
  2.10 Sample Scored Problem 24
  2.11 Directions for Printing 24
  2.12 CEDA 26
  2.13 The CEDA Translated into Other Languages 30
      2.13.1 Ukrainian 30
      2.13.2 Spanish 34
      2.13.3 Korean 38
      2.13.4 French 42

        2.13.5 Chinese (Mandarin).   46
        2.13.6 German   50
    References   54

3   **Future Directions: Uses**   57
    3.1   Importance to STEM   57
    3.2   Usability in Engineering Educational Programs.   58
    3.3   Use in Industry   61
    3.4   Use in NASA and the Military   65
    3.5   Conclusions   66
    References   67

About the Author   73

Abbreviations

APA	American Psychological Association
CEDA	Creative Engineering Design Assessment
CPS	Creative Personality Scale
CRT	Cognitive Risk Tolerance Survey
CSI	Creativity Support Index
CT	Creative Temperament Scale
EPM	Elementary Pragmatic Model
GPA	Grade Point Average
IDEO	Innovation Design Engineering Organization
NASA	National Aeronautics and Space Administration
NSF	National Science Foundation
PCT	Purdue Creativity Test
PVST-R	Purdue Visualization Spatial Test Rotations
SMOG	Simple Measure of Gobbledygook
STEM	Science, Technology, Engineering Mathematics
TLX	Task Load Index
TRIZ	an acronym in Russian that translates in English to Theory of Inventive Problem Solving
TTCT	Torrance Tests of Creative Thinking

Introduction

Creativity is no longer an optional accessory. Instead, creativity is a necessity for innovation and prosperity. Creativity is a resource that is needed to survive and solve everyday problems. Through practicing creative engineering design, we can benefit humankind, especially now in our current global economic climate. Creativity is a choice; one must desire to actively use their own creativity. One needs to seek out opportunities to use creativity. Like a muscle, creativity needs to be exercised on a regular basis. Just as one wishes to sing, play cello, piano or guitar well, one needs to practice their instrument regularly. Creative skills can be taught; however, one must be interested and engaged in using their own creativity as a lifestyle choice. Individual creativity is related to positive emotions and health which have traditionally been overlooked in psychology.

Engineering design has a long tradition of benefiting society and humankind from conveniences to technology. Many of our modern conveniences that we often take for granted stem from the creative engineering design process. The CEDA is a tool for assessing creative engineering design. The CEDA can also be used to assess the learning process and using creative engineering design skills.

Please be sure to read the book in entirety for adhering to purposes, goals and objectives as well as follow directions before implementing your uses of the CEDA. If you need further clarification, please contact Dr. Charyton.

Chapter 1
An Overview of the Relevance of Creative Engineering Design: Background

1.1 The Importance of Creativity in Engineering Design

Creativity research in engineering began to blossom in the 1950s (Ferguson 1992). The recommendations of Vannevar Bush, an electrical engineer from MIT, led to establishment of the National Science Foundation in 1950. At the same time, American Psychological Association (APA) President, J.P. Guilford, identified the need for creativity research (Guilford 1950). "The research design, although not essentially new, should be of interest" (Guilford, 1950, p. 444). Guilford (1950) elaborated, "A creative pattern is manifest in creative behavior, which includes such activities as inventing, designing, contriving, composing, and planning. People who exhibit these types of behavior to a marked degree are recognized as being creative" (p. 444). Guilford (1950) also stated that creative people have novel or new ideas.

In the early 1960s, the National Science Foundation (NSF) sponsored conferences on "scientific creativity." Yet, "as interest in engineering design faded in most engineering schools, creativity was put on a back burner" (Ferguson 1992, p. 57). More recently, creativity has received greater attention as a necessity, rather than an accessory in engineering design (Charyton and Merrill 2009). In today's times, creativity is appreciated even more as a vital resource.

Creativity and innovation have been a hallmark of engineering. "Creativity is certainly among the most important and pervasive of all human activities. Homes and offices are filled with furniture, appliances, and other conveniences that are products of human inventiveness" (Simonton 2000, p. 151). Psychology is valuable for addressing creativity in education by promoting learning through metacognition and self-reflective activities (Ishii and Miwa 2005).

The problems that engineers facing today demand original thinking. To remain competitive globally, engineering firms rely on creative individuals and creative teams to develop new products that drive the field forward. *Design News* (2007) reported that 65 percent of engineers in the workforce (from mechanical, application, and manufacturing engineering companies) agreed that today's engineers need to be more creative and innovative to be globally competitive

C. Charyton, *Creative Engineering Design Assessment*,
SpringerBriefs in Applied Sciences and Technology,
DOI: 10.1007/978-1-4471-5379-5_1, © The Author(s) 2014

(Christiaans and Venselaar 2005). The need for creativity in engineering has led to the development of numerous creativity support tools meant to aid engineers and designers in the creative design process (Baillie and Walker 1998).

The method of teaching creativity to engineering students has also been a key concern (Salter and Gann 2003). Engineering creativity specifically encompasses problem finding and problem solving. However, problem finding has not been a major focus of education. According to Nickerson (1999), creative problem finding offers another avenue for increasing creative production in engineering. Problem finding is common for an engineering designer who needs to think about and solve unforeseen problems (Ferguson 1992). An engineer's imagination and creativity have the power to develop technological solutions to problems (Deal 2001). This can be achieved through problem finding and problem solving.

Acceptance of creative ideas is also relevant. Csikszentmihalyi (1999) suggested that the person, field, and domain are relevant to understanding creativity and innovation. This theory postulates that acceptance of an idea, product, or process by the field, such as engineering and the domain (such as science or STEM—Science, Technology, Engineering, and Mathematics) is needed.

In engineering, there are often constraints in order to achieve design tasks. Engineers may increase creative production through understanding their domain-specific constraints. Stokes (2005) described how types of constraints (tasks, goals, subjects, functions, materials, and styles) may be unique to different domains. Research by Fink et al. (1996) found that the relative number of creative inventions increased significantly as the task became more highly constrained. Constraints on design may need further assessment as technology evolves (Mahboub et al. 2004; Redelinghuys 2001; Waks and Merdler 2003).

Engineers also utilize technology as a tool for design (Jones 1995; Smith 2002); however, human beings remain vital (McDougnel and Braungart 2002) for design and innovation. Products are sometimes invented by users to improve our everyday lives and social functioning (Von Hippel 2005). For example, communication tools that utilize technology often benefit consumers. Engineers can foresee these potential consumer needs. Such products can increase productivity and contribute to innovation which is the future of the engineering field (Jones 1995).

1.2 Creativity in Engineering Education

Today's engineers need to be creative and innovative in order to enhance creativity and innovation in the engineering field. Addressing these needs should begin in educational settings (Fleisig et al. 2009). The encouragement of creativity is vital in schools and curricula (Romeike 2006). However, education has the power to cultivate or stifle creativity (Burleson 2005). Since creativity is enhanced by confidence, self-development and positive mindset (Kang et al. 2011) educators play a key role in the learning process.

Creativity should be a vital part of engineering education as well as an important student outcome (Chiu and Salustri 2010). Engineering education is paramount for providing the nation with graduates who are innovative, creative, and critical in their thinking. These graduates are the human capital that will contribute to sustainability of the global economy (Yasin et al. 2009).

Empirical studies of learning methods incorporating educational and cognitive psychology have been successively implemented into engineering classes. Real-world applications, cooperative learning, active learning, and deductive and inductive learning are important for developing creativity (Felder et al. 2000). Reflection in action is vital to promote learning by doing (Schon 1983). Most importantly, students also need to practice design skills before they are assessed. Furthermore, experiential learning provides students opportunities to select assignments and promotes deeper learning.

Creative education enables students to make innovative products while promoting integrated cooperation (Ito et al. 2003). Creativity may be expressed as visualizing and manipulating images, greater openness to experience, and evaluating ideas (Hawlader and Poo 1989). Ito et al. (2003) suggest that imagination is attained by touching the concrete. Sulzbach (2007) notes a recent graduate emphasized teamwork, leadership, creative thinking, and problem solving: "no grades attached. That is the engineering student I want to hire."

Creativity and innovation are vital in most levels of engineering education, yet these topics are rarely expressed, investigated, or studied explicitly in coursework (Forbes 2008) despite the importance of creativity in engineering. Without training in the fundamentals of creativity, only 3 % of the population associate creativity with engineering (Stouffer et al. 2004). Ishii and Miwa (2004) found idea generation, idea embodiment, and collaboration of creative activities as important activities for learning. Therefore, projects with a keen personal interest may increase steps toward the commercialization of invention (Ruiz 2004). Furthermore, Jordan and Pereira (2009) found that sketching was valuable for teaching engineering design. We speculate that sketching is valuable toward the process of creative engineering design.

There has been a greater emphasis on enhancing creativity in upcoming engineers due to creativity's importance and value in the profession (Schmid 2006). The Council of Graduate Schools (2007) reported that the need to improve creativity and innovation in graduate students is of national importance in the United States. Engineering students recognize that creativity is important to learn. Approximately 40 years ago, a survey found that 87 % of engineering students agreed that creativity was a necessary skill for engineering (Gawain 1974). Furthermore, 77 % of engineering students stated that they would like to take a course in creativity and creative problem solving (Gawain, 1974).

In previous conversations with engineering faculty, a faculty member commented that the engineering curriculum has not been changed for over 200 years (B. Yantorno, personal communication, August, 24, 2003). Other engineering students, at Temple University and Ohio State University, as well as engineering students that have transferred from other universities have commented to me that they felt as though their classes only emphasized mathematics and not creativity. Another engineering faculty member stated that programs are quick to add more mathematics courses, yet reluctant to provide more creativity and innovation classes (D. Cropley, personal communication, August, 5, 2012).

In August 2012, Ohio State University's College of Engineering's Engineering Education Innovation Center (EEIC) first offered a student-initiated *Seminar for Promoting Creativity and Innovation* course. This course is being offered once per semester and is team-taught by faculty from various fields from all around the university including radiology, business, and psychology. This course beginning is an important "first step." Several students from this class spoke to me about my research and their interest to further updating the engineering curriculum beyond this course in order to engage more students in pursuing and maintaining their interests in engineering. Engineering and non-engineering majors at Ohio State University now have the opportunity to take a course directly in creativity and innovation as well as a course about the psychology of creativity. However, the need to enhance the engineering curriculum with creativity and innovation courses still exists.

Despite many approaches for teaching students about creativity, the engineering education system does not adequately prepare students for real-life problems in the professional world. One reason likely may be the education system's in-the-box thinking, which is not reflective of the real world (Tornkvist 1998). A second reason may be the fact that the pressures of real life can never be fully present in the classroom setting (Dallman et al. 2005). To address this, engineering educators have attempted to bring the world of professional engineering into the classroom. By introducing students to the more practical aspects of entrepreneurship and engineering, students learn by doing. Such practice better prepares students for the engineering problems they will encounter outside of the educational system in real-world settings (Cropley and Cropley 1998). In particular, simulation exercises via the learning-by-doing approach increase engineering skills and product development (Badran 2007) .

To stimulate "out-of-the-box" thinking, engineering educational programs have embraced interdisciplinary approaches in the design process. Interdisciplinary approaches have been especially beneficial in the field of engineering. Collaboration addresses the limitations of one discipline working alone as this may jeopardize design, construction, and use (Ghosh 1993).

At MIT, engineering students have been required to go outside of their discipline and learn new skills (Stouffer et al. 2004). MIT, however, is not the only engineering program emphasizing interdisciplinary studies. Educational programs in the College of Engineering at the University of Illinois at Urbana-Champaign, Stevens Institute of Technology, the School of Mechanical Engineering at the University of Leeds in England, and Arizona State University (Salter and Gann 2003) have also adopted an interdisciplinary approach.

At the University of Illinois at Urbana-Champaign, the College of Engineering incorporated different disciplines (the arts, humanities, and social sciences' departments) into their curriculum (Salter and Gann 2003). The Stevens Institute of Technology integrated engineering, architecture, computation, and product design in order to teach practical design problems (Salter and Gann, 2003). Oklahoma State University developed a course called *Strategies for Creative Problem Solving* (Mann and High 2003) in effort to increase retention (High et al. 2005). In 2008, Villanova University launched a *Creativity and Innovation* course that includes an exercise to act in different roles of thinking (problem finding, solution finding, object finding) for different colored hats known as DeBono's "Thinking Hats" as well as nonlinear problem-solving techniques.

The *Seminar on Promoting Creativity and Innovation* at Ohio State University, intended for engineering and non-engineering majors, is designed to provide students with the tools to refine their creative motivation as well as to encourage multidisciplinary innovation. By enabling students to explore the concept of creativity through a variety of guest speaker experiences, the course may foster a better appreciation for the processes of innovative design and prototyping. This promotes an environment of student-sustained innovation.

Enhancements in the educational curriculum are still needed at many institutions. Updates should include active and reflective learning, hands-on exercises, simulation exercises, and more courses addressing creativity and innovation.

1.3 Core Principles of Creative Engineering Design

Central themes specific to engineering creativity include *originality* (novelty) (Shah et al. 2003; Thompson and Lordan 1999; Weisberg 1999) and *usefulness* (applicability) (Larson et al. 1999; Shah et al. 2003; Thompson and Lordan 1999).

Engineers not only need to address esthetics like artists, but they also need to solve problems, prevent potential problems, and address utility within the constraints and parameters that have been designated. These aspects of creativity have been described as "functional creativity" (Cropley and Cropley 2005).

Functional creativity means that products designed by engineers typically serve a functional and useful purpose, unlike most typical fine art. Creative products emphasize novelty, resolution, elaboration, and synthesis (Cropley and Cropley 2005). Building on this, problem finding offers another avenue for increasing creative production (Nickerson 1999). Problem finding is a skill often found in and commonly associated with art, yet is also necessary in science and engineering. In art, problem finding often involves identifying social problems such as Picasso describing his dismay with the Spanish war in his artwork *Guernica* (Weisberg 1999).

Both *problem finding* and *problem solving* are relevant to an engineer's creativity; however, these attributes have not been specifically measured traditionally and in engineering creativity. Such attributes need to be assessed and further developed by appropriate educational intervention activities (Cropley and Cropley 2005).

These core principles were central to the development and theoretical framework of the Creative Engineering Design Assessment (CEDA) and are specifically measured in the CEDA tool.

1.4 Assessments to Measure Creativity in Engineering Design

To date, previous measures of engineering creativity have been limited. According to the literature available, only a few measures were developed to assess creative abilities in engineering design. These include the Owens Creativity Test (1960) and the Purdue Creativity Test (PCT) (1960). The Owens Creativity Test (Owens 1960) was developed to assess mechanical engineering design. Test takers list possible solutions to mechanical problems (divergent thinking). Reliability ranged from .38 to .91, while validity ranged from .60 to .72. Validity was determined via the testing of the engineers in mechanically related occupations. This assessment tool is out of print and is no longer used.

The PCT was developed by Lawshe and Harris (1960) as an engineering personnel test, as a method for identifying creative engineers and their occupational potential. Participants are instructed to list as many possible uses for one or two shapes that are provided. The PCT has adequate reliability (.86–.95) and modest validity (29–73 % for low scorers and high scorers, respectively). Validity was determined by assessing professional engineers (process and product engineers) working in industry. Participants are instructed to generate original and novel possible uses for single objects or pairs of objects. Scoring is based on fluency (number of uses) and flexibility (differing categories of uses). Although a reliable and valid measure, limitations include little use in the field of engineering. This assessment measures engineering creativity only by assessing fluency (number of responses) and flexibility (categories of responses) and does not directly assess originality. Both the Owens Creativity Test and the PCT are limited in that they only measure divergent thinking, list of potential uses, but not convergent thinking.

The CEDA offers a new method for assessing creative engineering design. Participants are asked to sketch designs that incorporate one or several three-dimensional objects, list potential users (people), as well as perform problem finding (generate alternative uses for their design) and problem solving in response to specific functional goals. Sketching is instrumental in designing problem solving (Goldschmidt and Smolkov 2006) and results in creative solutions. Some speculate that engineers think in pictures (Grandin 2006; B. Gustafson, personal communication, May 25, 2010). The sketching aspect of the CEDA is engineering specific and useful for spatial manipulations that are domain specific to engineering.

Creativity in psychology has traditionally emphasized divergent thinking skills (Torrance 1974; Guilford 1984). In the CEDA model, convergent science and divergent practices are integrated as necessary functions for creative engineering design. Schon (1983) reported that we have become aware of the importance of actual practice that encompasses uncertainty, complexity, and uniqueness in convergent science and divergent practices. Engineering often requires the need to solve problems in these types of ambiguous situations. However, deriving alternative solutions through problem finding is essential. Both *problem solving* and *problem finding* are important for creativity in engineering.

In order to be creative in engineering, solving problems is vital; however, determining when there is a problem to solve may be even more important (Ghosh 1993). Creativity support tools have focused on generating possible solutions, but not on identifying new problems (Baillie and Walker 1998). Yet, despite the importance of *problem finding* (identifying current problems or recognizing potential problems that may occur), the literature in engineering has traditionally been meager. This is true for assessing both engineering creativity and problem finding. To date, the CEDA is one of the only tools that assesses both *problem solving* and *problem finding* (Charyton et al. 2008).

The CEDA builds on and improves upon features of the PCT (Lawshe an Harris 1960) as well as Guilford's (1984) model of divergent thinking in that the questions are open-ended. The CEDA also assesses fluency (number of ideas), flexibility (categories of ideas, types of ideas, grouping of ideas), and originality (new ideas, novelty). However, the CEDA differs from the PCT in that it was not designed solely as a divergent thinking test. Furthermore, the CEDA was developed to specifically measure creativity unique to engineering design. Design is crucial toward creativity and innovation for *users* and customers (Cockton 2008).

Engineering creativity involves both *convergent thinking* (generating one correct answer) and *divergent thinking* (generating multiple responses or answers) (Charyton et al. 2008; Charyton and Merrill 2009). In the CEDA, students are asked to generate up to two novel designs to fulfill a generalized goal. The rationale for this limit is to work within the time *constraints* of the test and to elicit higher-quality responses. Also, because there are five steps to each design, the process requires more elements than just listing uses. In the CEDA, *divergent thinking* is assessed by generating multiple solutions. *Convergent thinking* is assessed by solving the problem posed. *Constraint satisfaction* is assessed by complying with the parameters of the directions and also adding additional *materials* and manipulating the objects as desired. *Problem finding* (identifying other potential problems) is assessed by identifying *other uses* for the design. *Problem solving* (finding a solution to a specific problem) is assessed by deriving a novel design to *solve the problem* posed.

In engineering, creativity requires originality, adaptiveness, problem solving (Weisberg 1986, 1999), and usefulness (Larson et al. 1999; Nickerson 1999).

Unlike previous measures, the revised CEDA also measures originality and usefulness, which, to date, is a unique component when compared to other general creativity and engineering creativity measures.

References

Badran, I. (2007). Enhancing creativity and innovation in engineering education. *European Journal of Engineering Education, 32*(5), 573–585.

Baillie, C., & Walker, P. (1998). Fostering creative thinking to student engineers. *European Journal of Engineering Education, 23*(1), 35.

Burleson, W. (2005). Developing creativity, motivation, and self-actualization with learning systems. *International Journal of Human-Computer Studies, 63*(4–5), 436–451.

Charyton, C., & Merrill, J. A. (2009). Assessing general creativity and creative engineering design in first year engineering students. *Journal of Engineering Education, 98*(2), 145–156.

Charyton, C., Jagacinski, R. J., & Merrill, J. A. (2008). CEDA: A research instrument for creative engineering design assessment. *Psychology of Aesthetics, Creativity, and the Arts, 2*(3), 147–154.

Christiaans, H., & Venselaar, K. (2005). Creativity in design engineering and the role of knowledge: Modelling the expert. *International Journal of Technology and Design Education, 15*(3), 217–236.

Chiu, I., & Salustri, F. A. (2010). Evaluating design project creativity in engineering design courses. *CEEA2010: Inaugural Conference of the Canadian Engineering Education Association* (pp. 1–6),

Cockton, G. (2008). Designing worth: Connecting preferred means to desired ends. *Interactions, 15*(4), 54–57.

Council of Graduate Schools Advisory Committee on Graduate Education and American Competitiveness. (2007). *Graduate education: The backbone of American competitiveness and innovation*. Washington, DC: Council of Graduate Schools.

Cropley, D., & Cropley, A. (1998). Teaching engineering students to be creative-program and outcomes. *Australasian Association of Engineering Education: 10th Annual Conference.*

Cropley, D. & Cropley, A. (2005). Engineering creativity: A systems concept of functional creativity. In J.C. Kaufman & J. Baer (Eds.), *Creativity across domains: Faces of the muse* (pp. 169–185). Mahwah, NJ: Lawrence Erlbaum Associates.

Csikszentmihalyi, M. (1999). Implications of a systems perspective for the study of creativity. In R. J. Sternberg (Ed.), *Handbook of Creativity*. Cambridge: MIT Press.

Dallman, S., Nguyen, L., Lamp, J., & Cybulski, J. (2005). Contextual factors which influence creativity in requirements engineering. *Proceedings of 13th European Conference on Information Systems ECIS.*

Deal, W. F. (2001). *Imagineering: Designing robots imaginatively and creatively*. The Technology Teacher, 17–25.

Felder, R. M., Woods, D. R., Stice, J. E., & Rugarcia, A. (2000). The future of engineering education. II. Teaching methods that work. *Chemical Engineering Education, 34*(1), 26–39.

Ferguson, (1992). *Engineering and the mind's eye*. Cambridge: The MIT Press.

Fink, R. A., Ward, T. B., & Smith, S. M. (1996). *Creative cognition: Theory, research and applications*. Cambridge: The MIT Press.

Fleisig, R., Mahler, H., & Mahalec, V. (2009). Engineering design in the creative age. *2009 ASEE Annual Conference and Exposition*, June 14–June 17 2009, Boeing.

Forbes, N. S. (2008). A module to foster engineering creativity: An interpolative design problem and an extrapolative research project. *CEE: Chemical Engineering Education, 42*(4), 166–172.

Forging a Path for Today's Engineers: More outsourcing, a more diversified workload and more money for engineers in 2007. (2007). *Design News, July 16, 2007*, 62. http://www.designnews.com/document.asp?doc_id=217539&dfpPParams=aid_217539&dfpLayout=article.

Gawain, T. H. (1974). Some reflections on education for creativity in engineering. *IEEE Transactions on Education, 17*(4), 189–192.

Ghosh, S. (1993). An exercise in inducing creativity in undergraduate engineering students through challenging examinations and open-ended design problems. *IEEE Transactions on Education, 36*(1), 113–119.

Goldschmidt, G., & Smolkov, M. (2006). Variances in the impact of visual stimuli on design problem solving performance. *Design Studies, 27,* 549–569.

Grandin, T. (2006). *Thinking in pictures: My life with autism.* New York: Random House, Inc.

Guilford, J. P. (1950). Creativity. *American Psychologist, 5,* 444–454.

Guilford, J. P. (1984). Varieties of divergent thinking. *Journal of Creative Behavior, 19,* 1–10.

Hawlader, M. N. A., & Poo, A. N. (1989). Development of creative and innovative talents of students. *The International Journal of Applied Engineering Education, 5*(3), 331–339.

High, K.A., Mann, C & Lawrence, B. (2005). Problem solving and creativity experiences for freshmen engineers. *Proceedings of the 2005 American Society of Engineering Education Annual Conference & Exposition.*

Ishii, N., & Miwa, K. (2004). Creativity education based on participants' reflective thinking on their creative processes. *Transactions of the Japanese Society for Artificial Intelligence, 19*(2), 126–135.

Ishii, N., & Miwa, K. (2005). Supporting reflective practice in creativity education. *Proceedings of the 5th conference on Creativity & Cognition, London, England.*

Ito, H., Fujimoto, H., Sakuramoto, I., Ishida, K., Kaneshige, A., Kadowaki, S., & Murakami, H. (2003). Practical education for students to enable to make a product creatively. *Proceedings of the International Conference on Agile Manufacturing, Advances in Agile Manufacturing, ICAM 2003* (pp. 645–650), December 4–6, 2003.

Jones, M. (1995). Company succeeds thanks to the creativity of its engineers. *Design News, 59.*

Jordan, S., & Pereira, N. (2009). Rube goldbergineering: Lessons in teaching engineering design to future engineers. *ASEE Annual Conference and Exposition,* June 14–17 2009, BOEING.

Kang, A., Chung, S., & Ku, J. (2011). A study on the effective lesson plan of creative engineering design education for the creativity improvement of the students of engineering college. *3rd International Mega-Conference on Future-Generation Information Technology, FGIT 2011, in Conjunction with GDC 2011* (pp. 108–119), December 8–10 2011, 7105 LNCS.

Larson, M. C., Thomas, B., & Leviness, P. O. (1999). Assessing the creativity of engineers. *Design Engineering Division: Successes in Engineering Design Education, Design Engineering, 102,* 1–6.

Lawshe, C. H. & Harris, D. H. (1960). *Manual of instructions to accompany Purdue Creativity Test forms G and H.* Princeton: Educational Testing Services.

Mahboub, K. C., Portillo, M. B., Liu, Y., & Chadranratna, S. (2004). Measuring and enhancing creativity. *European Journal of Engineering Education, 29*(3), 429–436.

Mann, C. & High, K.A. (2003). A pilot study for creativity in a freshman introduction to engineering course. *Proceedings of the 2003 American Society of Engineering Education Annual Conference & Exposition.*

McDougnel, W. & Braungart, N. (2002). *Cradle to cradle: Remaking the way we make things.* New York: North Point Press.

Nickerson, R. S. (1999). Enhancing creativity. In R.J. Sternberg (Ed.), *Handbook of creativity* (pp. 392–430). Cambridge: Cambridge University Press.

Owens, W. A. (1960). *The Owens creativity test.* Ames, IA: Iowa State University Press.

Redelinghuys, C. (2001). Proposed measures for invention gain an engineering design. *Journal of Engineering Design, 11,* 245–263.

Romeike, R. (2006). Creative students—what can we learn from them for teaching computer science? *6th Baltic Sea Conference on Computing Education Research—Koli Calling 2006* (Vol. 276, pp. 149–150), February 1, 2006.

Ruiz, F. (2004). Learning engineering as art: An invention center. *International Journal of Engineering Education, 20*(5), 809–819.

Salter, A., & Gann, D. (2003). Sources of ideas for innovation in engineering design. *Research Policy, 32*(8), 1309–1324.

Schon, D. A. (1983). *The reflective practitioner: How professionals think in action.* New York: Basic Books.

Schmid, K. (2006). A study on creativity in requirements engineering. *Softwaretechnik-Trends*, *26*(1), 1–2.

Shah, J. J., Smith, S. M., & Vargas-Hernandez, N. (2003). Metrics for measuring ideation effectiveness. *Design Studies, 24*(2), 111–134.

Simonton, D. K. (2000). Creativity: Cognitive, personal, developmental and social aspects. *American Psychologist, 55(1),* 151–158.

Smith, E. M. (2002). Teamwork with the next generation: A playful take on robots, draws students to technology, and gives engineers unexpected insights into their work and themselves. *Mechanical Engineering,* 46–49.

Stokes, P. D. (2005). *Creativity from constraints: The psychology of breakthrough.* NewYork: Springer.

Stouffer, W., Russell, J. S., & Oliva, M. G. (2004). Making the strange familiar: Creativity and the future of engineering education. *Proceedings of the 2004 American Society for Engineering Education Annual Conference & Exposition.*

Sulzbach, C. (2007). Enhancing engineering education? Concrete canoe competition., *114th Annual ASEE Conference and Exposition 2007,* June 24–27, 2007.

Thompson, G., & Lordan, M. (1999). A review of creativity principles applied to engineering design. *Proceedings of the Institution of Mechanical Engineers—Part E—Journal of Process Mechanical Engineering, 213*(1), 17–31.

Torrance, E. P. (1974). *Torrance tests of creative thinking.* Lexington: Personnel Press/Ginn and Co./Xerox Education Co.

Tornkvist, S. (1998). Creativity: Can it be taught? The case of engineering. *European Journal of Engineering Education, 23*(1), 5–12.

Von Hippel, E. (2005). *Democratizing innovation.* Cambridge: The MIT Press.

Waks, S., & Merdler, M. (2003). Creative thinking of practical engineering students during a design project. *Research in Science & Technological Education, 21*(1), 101–121.

Weisberg, R. W. (1999). Creativity and knowledge: A challenge to theories. In R.J. Sternberg (Ed.), *Handbook of creativity* (pp. 226–250). Cambridge: Cambridge University Press.

Weisberg, R. W. (1986). *Creativity: Genius and other myths.* New York, NY: W. H. Freeman and Company.

Yasin, R. M., Mustapha, R., & Zaharim, A. (2009). Promoting creativity through problem oriented project based learning in engineering education at Malaysian polytechnics: Issues and challenges. *8th WSEAS International Conference on Education and Educational Technology, EDU '09* (pp. 253–258), October 17–19, 2009.

Chapter 2
The CEDA Manual and Directions

2.1 Introduction

This manual includes directions for administering and scoring the Creative Engineering Design Assessment (CEDA). As previously suggested (Charyton and Merrill 2009) the CEDA is appropriate for use in high school through graduate school and could be easily used by engineering educators (Charyton et al. 2011). Further research is needed to see whether this tool is appropriate for engineers in industry. While future research will include this tool in industry, the CEDA has already been demonstrated to be related with another measure that was already normed on professional engineers in industry. Previous research has also shown that the CEDA is moderately related to other engineering creativity measures including the Purdue Creativity Test and the Purdue Spatial Visualization Test–Rotations. The Purdue Creativity Test was normed on product, process and project engineers working in industry with an average of 12 years experience from various fields of engineering including mechanical engineering. After establishing reliability, convergent validity, and discriminant validity, the CEDA tool is appropriate for dissemination.

The following resources are recommended in order to understand the terminology, theoretical framework and measuring components within the CEDA.

2.2 Recommended Resources

Charyton, C., Jagacinski, R.J., Merrill, J.A., *Clifton, W. & *Dedios, S. (October, 2011). Assessing engineering creativity and creative engineering design in first-year engineering students. *Journal of Engineering Education, 100,* 778–799.
 *indicates students supervised by Dr. Christine Charyton
 Link to pdf of article and You Tube Video: http://www.jee.org/2011/October/08

C. Charyton, *Creative Engineering Design Assessment,*
SpringerBriefs in Applied Sciences and Technology,
DOI: 10.1007/978-1-4471-5379-5_2, © The Author(s) 2014

Charyton and Merrill (2009). (also cited in Encyclopedia Britannica online retrieved November 18, 2009 from http://www.britannica.com/bps/additionalcontent/18/39246485/Assessing-General-Creativity-and-Creative-Engineering-Design-in-First-Year-Engineering-Students).

Charyton et al. (2008)

Charyton and Snelbecker (2007)

Charyton (2005)

Charyton (2008)

2.3 Directions for Administering the CEDA

Before administering the Creative Engineering Design Assessment (CEDA), the administrator should have a copy of the manual, read the materials and become familiar with all aspects of administration.

Participants are provided the following instructions:

At the top of the following, each page is a set of 2, 3, or 4 three-dimensional figures. Please use one or more of these figures to generate two original designs that will accomplish the general goal written below them. You can imagine that the figures are made of any material you wish and can be any size that you wish for each design. They can be solid or hollow and can be manipulated in any manner you wish. You may combine the figures on each page and may draw additional elements as required by your design. However, each figure can only be used once per design. On each page, be sure to:

1. Sketch your designs.
2. Label each design (provide a brief description—what is your design?).
3. Describe your materials.
4. Identify additional problems that your design may solve.
5. Identify the users (specific persons) of each design.

Total time for this assessment is 30 min for 3 pages, or about 10 min per page. You may use your time as you see fit. Two designs should be created per page. Additionally, at least one response should be indicated for each of the boxes below your sketch for each design. You may use a pen or pencil, whichever you prefer.

Instruct the participants to print their name or identifying number on the face sheet of the assessment. There are a total of three problems for the revised version of the CEDA. Use for the CEDA requires responses that are created and sketched through a design process that is unique to the individual. The total working time on the CEDA is approximately 30 min with 10 min allotted per problem as described in the directions above. Participants should be instructed to complete the CEDA in this manner with time constraints; however, administration may take place outside of the administration environment. Participants are instructed to generate two designs for each problem.

2.4   Theoretical Framework of the CEDA

Figure 2.1 describes how each item on the CEDA addresses these theoretical constructs. *Divergent thinking* is assessed by generating multiple solutions to the problem. *Convergent thinking* is assessed by solving the problem posed by creating at least three designs for three problems. *Constraint satisfaction* is assessed by complying with the parameters of the directions and also adding additional materials and manipulating the objects as desired. If items are duplicated, the design is disqualified (DQ). Designs are also DQ if they are not *appropriate* responses to solve the problem. *Problem finding* is assessed by identifying other uses for the design. This is the most difficult aspect of the entire problem. *Problem solving* is assessed by deriving a novel design to solve the problem posed. This means solving the problem appropriately yet in a novel manner.

A readability and comprehension analysis was conducted on the CEDA to determine the appropriateness for college students. The analysis measure known as the Simple Measure of Gobbledygook (SMOG) (McLaughlin, 1969), an online program, was used to assess the reading and comprehension level of the CEDA, available at:

http://www.harrymclaughlin.com/SMOG.htm

The SMOG formula utilizes established readability formulas that match scores with the actual education level. The online SMOG calculator uses McLaughlin's formula yielding a 0.985 correlation with the grade level of readers having 100 % comprehension of the tested materials.

The SMOG is designed for evaluating the reading level of materials that can be read independently by a person without assistance from a teacher or instructor (Richardson and Morgan 1990). Readability is recommended at the 6–8th grade level for educational materials for the general public. The SMOG Grade for the CEDA was 8.81, being the 8th grade level, equivalent to a junior high school, which relates to a newspaper reading comprehension level. Therefore, the CEDA

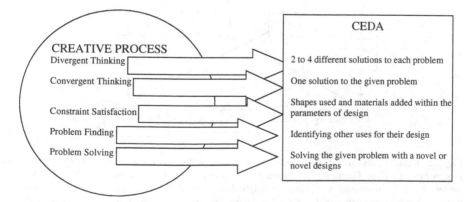

Fig. 2.1 Creative engineering design assessment meta-cognitive processes measured

would also be appropriate and useful for precollege students at the junior high and high school levels.

2.5 Components of Scoring

Figure 2.2 depicts the correlations among the components of the CEDA. The strong correlation ($r = 0.86$) between Fluency (number of ideas) and Flexibility (types of categories of ideas) reflects that both measure divergent thinking in terms of number of designs, although Flexibility uses more categorical analysis. Given this greater abstraction for Flexibility, it is perhaps not surprising that its correlations with Originality (novelty of ideas) ($r = 0.58$) and Usefulness (practicality for potential or current uses as well as number of potential uses) ($r = 0.46$) are numerically higher than those for Fluency ($r = 0.46$ and $r = 0.39$, respectively). The relatively high correlation between Originality (new ideas) and Usefulness (practicality) ($r = 0.65$) is perhaps surprising given that these are distinct constructs that are both central to engineering creativity. However, their relationship may be higher in an engineering population, which values both Originality and Usefulness more than other fields or domains.

The inter-rater reliabilities between the four engineering judges and the one psychology judge were calculated for each of the components, except Fluency, which simply consisted of a count of items. The reliabilities for Flexibility ($r = 0.83$, $p < 0.01$), Originality ($r = 0.59$, $p < 0.01$), and Usefulness ($r = 0.46$, $p < 0.01$) were lower than the inter-rater reliability of the overall CEDA scores without Usefulness ($r = 0.92, p < 0.01$) and with Usefulness included ($r = 0.81, p < 0.01$).

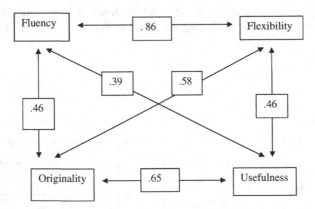

Fig. 2.2 Correlations among the components assessed within the Creative Engineering Design Assessment (CEDA). Fluency is the amount of responses. Flexibility is the amount of types or categories of the responses per problem. Originality is novelty or new ideas that are assessed based on a rubric consisting of descriptors and numbers on a scale from 0 to 10. Usefulness is the practicality of the design for current and/or potential future uses on a Likert's scale from 0 to 4

The magnitude of the reliabilities indicates the difficulty of judging Originality and Usefulness. The higher reliability of the overall CEDA scores is based on a combination of these component measures. The CEDA is comparable with the Purdue Creativity Test (PCT) on Fluency and Flexibility.

We also assessed inter-rater reliability for 60 randomly selected PCT's, 30 from Fundamentals of Engineering I, and 30 from Fundamental of Engineering II. Reliability was comparable with the CEDA for Flexibility ($r = 0.89$, $p < 0.01$) and overall PCT scores ($r = 0.93$, $p < 0.01$) in our study. However, the CEDA contributes to existing measures by assessing essential components of creative engineering design, such as Originality and Usefulness, which are core features of creativity, that are especially important in engineering yet have not been previously directly measured (Charyton et al. 2011).

2.6 Properties of the CEDA

2.6.1 Reliability

Previous correlations among the CEDA scores of two judges were conducted to identify their relationships with each other and establish reliability. Judges were in agreement ($r = 0.98$) with their overall scoring. Inter-rater reliability was high for Flexibility ($r = 0.90$ and $r = 0.98$) and Originality ($r = 0.80$ and $r = 0.85$) indicating consistency in both test and retest measures, respectively (Charyton and Merrill 2009).

The CEDA was consistent for test and retest reliability ($r = 0.56$) like the other general creativity measures such as the Creative Personality Scale (CPS) ($r = 0.57$), Creative Temperament Scale (CT) ($r = 0.51$), and Cognitive Risk Tolerance Scale (CRT) ($r = 0.43$), ($p < 0.01$) for all comparisons (Charyton and Merrill 2009).

Reliability was important to re-establish since we modified the CEDA to assess Usefulness in addition to Originality. The four engineering judges were in agreement with the psychology judge at the following levels ($r = 0.95$), ($r = 0.88$), ($r = 0.91$), and ($r = 0.93$), $p < 0.01$ for all comparisons (Charyton et al. 2011).

2.6.2 Validity

2.6.2.1 Discriminant Validity

In a previous study, Charyton and Snelbecker (2007) found that a music improvisation creativity measure was not related to general creativity constructs

(CPS, CT, and CRT), yet the Purdue Creativity Test (PCT) demonstrated a modest relationship with these general creativity measures. The CEDA demonstrated discriminant validity from these other general creativity measures (Charyton et al. 2008), like the domain specific music improvisation creativity measure had in previous studies (Charyton 2005; Charyton 2008; Charyton and Snelbecker 2007). The general creativity measures are described as follows:

CPS: Creative Personality Scale. The CPS of the Adjective Checklist (ACL) (Gough 1979) was previously administered to assess creativity attributes. According to Gough (1979), aesthetic dispositions are related to creative potential. This instrument was designed as an appraisal of the self. This test was selected because it is highly regarded, reliable and widely used as a general creativity test (Plucker and Renzulli 1999; Oldham and Cummings 1996).

CT: Creative Temperament Scale. The CT (Gough 1992) was adapted from the California Psychological Inventory (CPI), which was designed to assess personality characteristics and predict what people will say and do in specific contexts. Gough (1992) suggested that this measure is capable of forecasting creative attainment in various domains, both within and outside of psychology. Any domain requires skills specific to the domain, yet this measure assesses general personality qualities cutting across disciplines. The Creative Temperament Scale is one of the special purpose scales of the CPI.

CRT: Cognitive Risk Tolerance Survey. The CRT (Snelbecker et al. 2001) consists of 35 self-report items to assess an individual's ability to formulate and express one's ideas despite potential opposition. Responses are on a Likert's Scale ranging from 0 (Very Strongly Disagree) to 9 (Very Strongly Agree). Higher scores indicate higher levels of cognitive risk tolerance. The Cognitive Risk Tolerance Survey was developed as an extension of an earlier risk tolerance model developed by Snelbecker and colleagues (Roszkowski et al. 1989; Snelbecker et al. 1990). Charyton and Snelbecker (2007) found the CRT measure was moderately correlated with the CPS ($r = .36, p < 0.01$) and CT ($r = .34, p < 0.01$), which were moderately related to each other ($r = .35, p < 0.01$). CRT may be a component of general creativity that is moderately related to, yet different from other general creativity measures. This measure was selected as a distinct component of general creativity.

Discriminant validity for the CEDA was established with general creativity measures, respectively ($r = -0.01$ (CPS); $r = -0.13$ (CT), $r = -0.19$ (CRT), $p > 0.05$) suggesting that the CEDA is domain specific to engineering.

2.6.2.2 Convergent Validity

Correlations between the CEDA and other engineering creativity and spatial measures were conducted to establish convergent validity. The CEDA was moderately correlated with the PCT ($r = 0.39, p < 0.01$) and slightly correlated with the Purdue Visualization Spatial Test–Rotations (PVST–R) ($r = 0.19, p < 0.05$). The CEDA, including usefulness in the formula of assessment, demonstrated similar results. The CEDA with usefulness was moderately correlated with the

PCT ($r = 0.31$, $p < 0.01$) and slightly correlated with the PVST–R ($r = 0.21$, $p < 0.05$). These findings suggest that creative engineering design overlaps with spatial skills. This finding is logical since sketching requires spatial skills. Furthermore, in the CEDA, participants are instructed to manipulate the objects in any manner they desire without replication. In the PVST–R, participants are instructed to rotate the objects.

The relationship among the variables is consistent with the previous (or initial) CEDA scoring formula and the revised CEDA scoring formula. Figure 2.3 depicts the initial CEDA scoring and the revised CEDA scoring (new CEDA scoring in parentheses) that includes Usefulness. In this study, the other domain-specific engineering specific measures are moderately related to the CEDA, demonstrating that the CEDA is more like other engineering creativity measures (PCT) and engineering spatial measures (PVST–R). Both are domain specific to engineering. This contrasts with previous findings demonstrating that the CEDA was not like other general creativity measures. Thus, by directly assessing Originality and Usefulness, the CEDA assesses creativity as a well-accepted standardized definition that is also domain specific to engineering (Charyton et al. 2011).

The Purdue Spatial Visualization Test-Rotations is the most common test of engineering students' spatial visualization (Carter et al. 1987). The PSVT–R consists of 30 unfamiliar objects that the observer mentally rotates and has been used for first-year engineering programs (Bodner and Guay 1997). The PSVT–R was devised to test spatial development while minimizing analytic processing (Guay 1980). Using lines and symbols to represent thoughts and ideas of engineers can be more effective than purely verbal descriptions (Scribner and Anderson, 2005). The PSVT–R correlated significantly with participants' scores on spatial tasks (Kovac 1989). Males have previously performed better on PSVT questions than females (Guay 1978; Kinsey et al. 2007); however, the two genders scored equally on self-efficacy (belief in their own capabilities in order to accomplish the task), while upperclass students scored higher on both (Kinsey et al. 2007). The PSVT–R is often administered to freshmen with a course objective of assessing the spatial skills needed to succeed in subsequent engineering design graphics courses (Sorby and Baartmans 1996).

The PSVT–R has high construct validity for spatial visualization ability (Branoff 2000; Guay 1980). Guay reports internal consistency (reliability) (Kuder Richardson $r = 0.87$, 0.89 and 0.92) from 217 university students, 51 skilled machinists, and 101 university students, respectively (Guay 1980). Other studies also report high reliability (Kuder Richardson $r = 0.80$ or higher) (Branoff 2000; Scribner and Anderson 2005). Based on these analyses, most researchers agree that the PSVT is a good measure of spatial ability (Branoff 2000; Yue and Chen 2001).

2.6.3 Conceptualization

See Fig. 2.3.

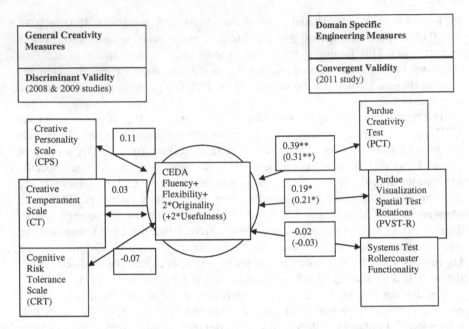

Fig. 2.3 Conceptualization of the relationship of the Creative Engineering Design Assessment (CEDA) with general creativity measures (2008; 2009) and domain-specific engineering measures (2011). The *top left* portion of the figure is based on previous research (Charyton et al. 2008), and the *top right* portion is based on current data in this study with the original and revised CEDA scoring formula. The revised scoring CEDA formula is in parentheses. The correlations for the revised formula with usefulness (2* usefulness added to the original CEDA formula) illustrates similar findings with the new scoring of the revised CEDA compared with the previous scoring method without usefulness. ** indicates statistical significance at the 0.01 level. * indicates statistical significance at the 0.05 level

2.7 How to Score the CEDA

See Table 2.1

2.8 Directions

For Fluency, Flexibility, Originality, and Usefulness, each question of the CEDA problems is described in regard to each CEDA question and revised CEDA scoring sheet. There are five parts for Fluency and Flexibility, eleven parts for Originality and five parts for Usefulness. The scoring sheet terms are provided within parentheses for scoring dimensions of the CEDA (Fluency: number, total designs, descriptions, materials, problems solved, and users); (Flexibility: Types or Classifications, total designs, descriptions, materials, problems solved, and users);

Table 2.1 The Scoring Sheet

Revised CEDA scoring sheet	Participant Number		Judge's name:
Fluency (number)	*Problem 1*	*Problem 2*	*Problem 3*
Fluency (total designs):			
Fluency (descriptions provided)			
Fluency (materials)			
Fluency (problems solved)			
Fluency (users)			
Judges begin here:			
Flexibility (types or classifications)			
Flexibility (Total designs)			
Flexibility (descriptions)			
Flexibility (materials)			
Flexibility (Problems solved)			
Flexibility (Users)			
Uniqueness/originality			
(assign 3 numbers per problem: Design 1,2 & Overall in sequential order)			
(1–10) per design	*D1 D2 Overall*	*D1 D2 Overall*	*D1 D2 Overall*
(0 dull			
1 commonplace,			
2 somewhat interesting,			
3 interesting			
4 very interesting			
5 insightful			
6 unique and different			
7 exceptional			
8 innovative			
9 valuable and beneficial to the field			
10 genius)			
Usefulness			
(assign 3 numbers per problem: Design 1, 2 & Overall in sequential order)			
	D1 D2 Overall	*D1 D2 Overall*	*D1 D2 Overall*
0 not useful			
1 somewhat useful			
2 moderately useful			
3 very useful			
4 indispensable			
Definition of usefulness: practicality of a design based on reliability, number of purposes, and number of occasions for application. The usefulness of a design can involve present uses and new uses in the future.			

Uniqueness/Originality (per design and overall) (0 dull, 1 commonplace, 2 somewhat interesting, 3 interesting, 4 very interesting, 5 insightful, 6 unique and different, 7 exceptional, 8 innovative, 9 valuable and beneficial to the field, 10 genius); Usefulness (per design and overall): practicality of a design based on reliability, number of purposes, and occasions for application.

The Usefulness of a design can involve present uses and new uses in the future (0 not useful, 1 somewhat useful, 2 moderately useful, 3 very useful, 4 indispensable). Usefulness would also account for sustainability. Greater sustainability would increase potential uses, purposes, and occasions for application.

2.8.1 Fluency

Participants taking the CEDA tool will most likely have two designs completed per problem. Designs may be similar or different from each other. To begin scoring the CEDA designs, all items answered should be counted. Fluency is the total number of designs. Fluency is counted for the Sketch, Description, Materials, Additional Problems Solved, and Users.

Fluency (Number) is a straightforward count for sketches (Total Designs) and description (Descriptions Provided). However, many participants select multiple materials (Materials) to create their designs. Therefore, we recommend checking carefully and underlining each material. For additional problems solved (Problems Solved), there may be more than one additional problem solved stated, so a thorough check is important to identify multiple responses. Last, many participants also select multiple users for each design and problem; therefore, users (Users) should also be thoroughly checked for multiple responses. Test administrators may also want to underline each User to identify multiple users.

However, when calculating **Fluency** for Users, if a participant writes **everyone, anyone, everybody**, and other related words the design should only receive a score of **one**, if there are no other responses. If other responses are also indicated, these **vague responses** would count as **zero** toward Fluency.

After the Fluency is calculated for all participants' design problems, begin scoring Flexibility, Originality, and Usefulness.

2.8.2 Flexibility

Flexibility is the number of categories, types, or classifications of responses. Are responses the same or different? If responses are the same, fewer points are assigned. For example, if we were to have participants' list categories of foods and if one participant responded as **orange, apple, and banana**, then they will only receive **one** point for flexibility since the items indicate only one type of category (**fruit**). If responses are different from each other, more points are assigned, for example, if another participant said **steak, apple, and bread**, that would count as **three** points for each different category (meat, fruit, and bread).

To score Flexibility (Total Designs) for the sketches look at both Designs. Are they the same? Or different? For example, in designs, which produce sound, are the cylinder and sphere in the same combination together (visually)? If they are in the same combination, they would receive one point for Flexibility. If the combinations are different, or if each design uses a different shape, then the problem would receive two points for Flexibility (different categories).

To score Flexibility (Descriptions) for the each description, determine, are they the same or different? Different descriptions receive two points, while the same description would receive one point (often times there are multiple descriptions).

To score Flexibility (Materials), be sure to check each for repetition or same idea. It is important to note that when participants have duplicate answers such as **metal, metal speaker, and rubber,** the problem would receive **three** points for **Fluency**, but only **two** points for **Flexibility**.

To score Flexibility (Problems Solved) for additional problems solved look for similar or different answers. For each different response, an additional point is given. Participants would often receive two points or more depending on the amount of responses. However, similar ideas would receive one point.

To score Flexibility (Users), check for duplication and categories of responses. Many participants have multiple responses. For example, if the participant responded **children** and **kids,** the problem would receive **one** point for Flexibility. If a participant responded **children, astronauts, and business people,** they would receive **three** points for both Fluency and Flexibility. However, if a participant responded **children, kids, and astronauts,** they would only receive two points for Flexibility while receiving three points for Fluency. Vague responses such as "everyone," "anyone," or "everybody" would receive only one point, if these are the only responses. If a participant responded **children, kids, anyone,** they would only receive one point for Flexibility. This is because children and kids are the same. Also, **anyone** does not gain any points. Anyone is too general and not thoughtful in the design process.

2.8.3 Originality

Before scoring Originality, the scorer or judge should read the scoring rubric for Originality. Originality is defined as novelty, new ideas, or unique ideas.

To score Originality (Uniqueness), rate each design on the scale from 0 to 10. Scorers or judges should think of a word on your own that describes each design and then look on the rubric list to find the word and assign that number to the design. Each problem has two designs. Each design should be assessed separately (D1, D2). Then, an overall evaluation of the entire problem should be rated. The Originality score for the entire problem (Overall) will be the score that is analyzed and becomes the overall Originality score for the problem. Although each design score can be inputted and analyzed, we recommend using the overall problem score. It is also important to note that this process of scoring each design is pertinent toward making an assessment of the overall Originality score per problem.

2.8.4 Usefulness

The definition of Usefulness is the practicality of the design, based on reliability, number of purposes, and number of occasions for application. The usefulness can involve present uses and new uses in the future.

To score Usefulness, rate each design on a scale from 0 to 4. This is a Likert's Scale where 0 is not useful, and 4 is indispensable. Like Originality, use a word that describes the usefulness of each design. It is important to keep in mind that usefulness incorporates the number of uses, present uses, and future uses. Be sure to rate Usefulness for each design (D1, D2) and then rate the overall usefulness per problem (Overall). Although each design score can be inputted and analyzed, it is recommended to use the overall problem score. It is also important to note that this process of scoring each design is pertinent toward making an assessment of the overall Usefulness score per problem.

It is important to note that if a participant does not follow the directions and duplicates a shape, then their design is disqualified and should not be assessed. However, participants can manipulate the shape or shapes and add materials as they see fit. If duplication appears to have been manipulated on a smaller scale, then the design can still be counted. For example, if five cylinders appear to be the same size and have been duplicated, then the design is disqualified (DQ). However, if the cylinder was manipulated into five smaller cylinders that appear to have been generated and manipulated from one cylinder, then the design can still be scored since the participant was successfully following the directions by not replicating objects. DQ can be assigned to each design where the directions were not followed properly, the response is not *appropriate* for the problem and when one or more objects have been duplicated. Remember items can be manipulated, but not duplicated. This is a judgment call by the rater. If there are two designs and one design is disqualified, the overall score would be the same as the design that was suitable for scoring. There are cases where both designs were disqualified. DQ should be notated when reporting results.

2.9 Sample Design Problem

See Table 2.2

Table 2.2 Sample Design Problem

Designs that produce sound.

Sketch	1	2
Description (What is your design?)	The individual uses the "stick - like" piece to talk into, almost like a microphone and the circle is a portable speaker	My design is a ball with the stick running through the middle of the ball. It makes noises when bounced
Describe the Materials	The circle is a speaker and the stick is metal with a receptor on the one end about 2 inches into the tube	The ball is made of rubber and filled with air so it can bounce, the stick is made from soft foam that projects the sound made from the speaker inside it
Additional Problems solved	The tube is better than a mic because you talk directly into it and it will pick up less surrounding sounds	A soft ball and soft speaker covering is a good safety and entertainment feature for kids
Users (persons that could use your design)	People in a conference room for a business meeting. Professors of a lecture	Children at young ages would love a bouncy ball that makes noises

2.10 Sample Scored Problem

See Table 2.3

2.11 Directions for Printing

Print the CEDA in black and white ink and photocopy for administration. Scoring Sheets should also be used to score all CEDAs and be printed in black and white hardcopy also. We recommend using hardcopies versus electronic copies for both the CEDA and Scoring Sheets.

Table 2.3 Sample Scored Problem

Revised CEDA Scoring Sheet	Participant Number		Judge's name:
Fluency (number)	*Problem 1*	*Problem 2*	*Problem 3*
Fluency (Total Designs):	2		
Fluency (descriptions provided)	2		
Fluency (materials)	7		
Fluency (Problems solved)	4		
Fluency (Users)	3		
Judges begin here:			
Flexibility (types or classifications)			
Flexibility (total designs)	2		
Flexibility (descriptions)	2		
Flexibility (materials)	6		
Flexibility (problems solved)	4		
Flexibility (users)	3		
Uniqueness/originality			
(assign 3 numbers per problem: Design 1,2 & Overall in sequential order)			
(1–10) per design	*D1 D2 Overall*	*D1 D2 Overall*	*D1 D2 Overall*
(0 dull			
1 commonplace,			
2 somewhat interesting,			
3 interesting			
4 very interesting	4		
5 insightful	5 5		
6 unique and different			
7 exceptional			
8 innovative			
9 valuable and beneficial to the field			
10 genius)			
Usefulness			
(assign 3 number per problem: Design 1, 2 & Overall in sequential order)			
	D1 D2 Overall	*D1 D2 Overall*	*D1 D2 Overall*
0 not useful			
1 somewhat useful			
2 moderately useful	2		
3 very useful	3 3		
4 indispensable			
Definition of usefulness: practicality of a design based on reliability, number of purposes, and number of occasions for application. The usefulness of a design can involve present uses and new uses in the future			

2.12 CEDA

Before beginning, please provide the following 3 pieces of identification:
  Course number _____ **Semester/Quarter:**_____
  Assigned Student Number _____
  CEDA: Creative Engineering Design Assessment.

At the top of the following, each page is a set of 2, 3, or 4 three-dimensional figures. Please use one or more of these figures to generate two **original** designs that will accomplish the general goal written below them. You can imagine that the figures are made of any material you wish and can be any size that you wish for each design. They can be solid or hollow and can be manipulated in any manner you wish. You may combine the figures on each page and may draw additional elements as required by your design. However, each figure can only be used once per design. On each page, be sure to:

1. Sketch your designs.
2. Label each design (provide a brief description—what is your design?).
3. Describe your materials.
4. Identify additional problems that your design may solve.
5. Identify the users (specific persons) of each design.

Total time for this assessment is 30 min for 3 pages, or about 10 min per page. You may use your time as you see fit. Two designs should be created per page. Additionally, at least one response should be indicated for each of the boxes below your sketch for each design. You may use a pen or pencil, whichever you prefer.

Designs that produce sound.

Sketch	1	2
Description (What is your design?)		
Describe the Materials		
Additional Problems solved		
Users (persons that could use your design)		

Designs that facilitate communication.

Sketch	1	2
Description (What is your design?)		
Describe the Materials		
Additional Problems solved		
Users (persons that could use your design)		

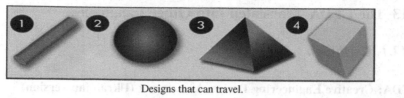

Designs that can travel.

Sketch	1	2
Description (What is your design?)		
Describe the Materials		
Additional Problems solved		
Users (persons that could use your design)		

2.13 The CEDA Translated into Other Languages

2.13.1 Ukrainian

CEDA: Creative Engineering Design Assessment (Ukrainian version)

До приступання виповнення анкети, прошу подати:

Число курсу: _____ Зазначення підвідділу: _____
Реєстреційне число студента: _____

Зверху кожного листка найдете 2-, 3- або 4-вимірні фігури.
Користуючися цими фігурами, проситься створити два оригінальні дезайни, які
виконюють загальну мету описану Вами знизу. Ви можете уявити, що ці фігури
можуть бути виконані з будь-якого матеріялу і можуть мати будь-який розмір. Вони
можуть бути пусті або вагомі, і можна їх маніпулювати в будь-який спосіб. Ви
можете поєднати фігури на одному аркуші і завгодно добавити додаткові елементи,
щоб виконати намічену ціль. Одне застереження: кожна фігура може бути
використана тільки одноразово в окремих дезайнах.

Проситься виповнити кожний аркуш наступними деталями:

1. Зарисовка дезайну.
2. Назва кожного дезайну з коротким описом, що в ньому представлено.
3. Опис матеріялу.
4. Намічені проблеми адресовані Вашим дезайном: (які певні питання Ваш дезайн
задовольняє?)
5. Яка користь Вашому дезайні? Хто може ним користуватися?

Проект має три окремі сторінки і зарахований на 30 хвилин--10 хвилин за аркуш.
Користуйтесь часом як Вам завгодно. На кожному аркуші повинно бути 2 дезайни і
для кожного дезайну, прошу відповісти на відокремлені питання. Ви можете
користуватися ручкою або олівцем, як найзручніше.

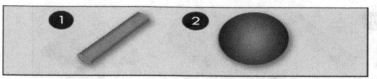

Дезайни, які творять ЗВУК

Рисунок	1	2
Опис (що представлено?)		
Опис матеріялів		
Додаткові питання використані дезайном		
Хто корустиватиметься дезайном?		

Дезайни, які сприяють КОМУНІКАЦІЇ

Рисунок	1	2
Опис (що представлено?)		
Опис матеріялів		
Додаткові питання використані дезайном		
Хто корустиватиметься дезайном?		

Дезайни, які ПОДОРОЖУЮТЬ

Рисунок	1	2
Опис (що представлено?)		
Опис матеріялів		
Додаткові питання використані дезайном		
Хто корустиватиметься дезайном?		

2.13.2 Spanish

Previo a comenzar, favor de proveer las 3 próximas piezas de identificación:

Numero de curso_____Letra de Sección_____Numero de Estudiante Asignado_____

ECDI: Examinación de Creatividad de Diseño en Ingeniería.

A lo alto de cada una de las siguientes, cada pagina es un conjunto de 2, 3, o 4 figuras de tres-dimensión. Favor de utilizar una o mas de estas figuras, previamente mencionadas, para generar dos diseños **originales** con el fin de lograr la meta escrita bajo las mismas. Las figuras pueden tomar cualquier tamaño y ser compuestas del material que desees en cada diseño; pueden ser solidas o huecas y ser manipuladas en cualquier manera que desees. Puedes combinar las figuras en cada pagina y dibujar elementos adicional como sean necesarios por tu diseño. Sin embargo, cada figura tiene el limite de ser utilizada solamente una vez, por diseño. En cada pagina, asegúrate de:

1. Dibujar tus diseños.
2. Proveer escritura de etiqueta (dar una descripción breve – ¿que es tu diseño?)
3. Describir tus materiales.
4. Identificar problemas adicionales que tu diseño puede resolver.
5. Identificar los usuarios (personas especificas) de cada diseño.

Tiempo total para este examen es 30 minutos para 3 paginas, o aproximadamente 10 minutos por pagina. Puedes utilizar el tiempo como veas necesario. Dos diseños necesitan ser creados por pagina. Además, es necesario indicar por lo menos una respuesta en cada una de las cajas bajo tu dibujo, por cada diseño. Puedes escoger entre lápiz o lapicero, como prefieras.

Diseños que producen sonido.

Dibujo	1	2
Descripción (¿Qué es tu diseño?)		
Describe tus Materiales		
Problemas adicionales resueltos		
Usuarios (que pueden sacar provecho del diseño)		

Diseños que faciliten la comunicación.

Dibujo	1	2
Descripción (¿Qué es tu diseño?)		
Describe tus Materiales		
Problemas adicionales resueltos		
Usuarios (que pueden sacar provecho del diseño)		

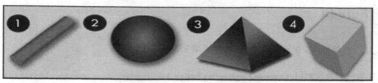

Diseños que pueden viajar.

Dibujo	1	2
Descripción (¿Qué es tu diseño?)		
Describe tus Materiales		
Problemas adicionales resueltos		
Usuarios (que pueden sacar provecho del diseño)		

2.13.3 Korean

과목 번호_____ 섹션(/학기) 정보 _____ 학생 식별 번호 _____

창의적 조합 능력 평가 (CEDA: Creative Engineering Design Assessment)

매 페이지의 상단에는 2차원, 3차원, 혹은 4차원의 도형들이 제시됩니다. 이 도형 중 한가지 이상을 이용하여 제시된 목적을 달성할 수 있는 **독창적인** 디자인 두 가지를 그려주십시오. 제시된 도형의 재질이나 크기는 귀하가 그리고자 하는 디자인에 따라 얼마든지 다르게 상상할 수 있습니다. 필요에 따라 도형의 속이 가득 차거나 비어있다고 상상할 수 있으며, 자유롭게 변형할 수도 있습니다. 각 페이지에 제시된 도형을 조합한 뒤 필요한 도형을 추가적으로 그릴 수도 있으나, 각 도형은 한 번씩만 이용할 수 있습니다. 매 페이지에 다음 항목을 그리거나 작성하여 주십시오.

1. 디자인의 스케치

2. 디자인의 명칭 (디자인의 용도를 간단히 설명)

3. 디자인에 이용된 도형의 재질

4. 디자인의 추가적인 용도 명세

5. 디자인을 이용할 수 있는 특정 인물 명세

본 평가에 소요되는 시간은 3 페이지 당 30분, 혹은 페이지당 10분 입니다. 필요하신 시간을 모두 이용하셔도 좋습니다. 한 페이지 당 두 가지의 디자인이 제시되어야 합니다. 다섯 개의 모든 문항에 대해 최소한 하나 이상의 답을 하셔야 합니다. 연필이나 펜 중 선호하는 것을 이용할 수 있습니다.

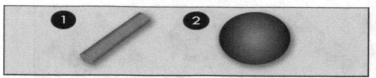

소리를 내는 물건을 만드십시오.

스케치	1	2
디자인의 명칭 (디자인의 용도를 간단히 설명)		
디자인에 이용된 도형의 재질		
디자인의 추가적인 용도		
디자인을 이용할 수 있는 특정 인물		

의사소통을 돕는 물건을 만드십시오.

스케치	1	2
디자인의 명칭 (디자인의 용도를 간단히 설명)		
디자인에 이용된 도형의 재질		
디자인의 추가적인 용도		
디자인을 이용할 수 있는 특정 인물		

이동할 수 있는 물건을 만드십시오.

스케치	1	2
디자인의 명칭 (디자인의 용도를 간단히 설명)		
디자인에 이용된 도형의 재질		
디자인의 추가적인 용도		
디자인을 이용할 수 있는 특정 인물		

2.13.4 French

Avant de commencer, veuillez donner trois pièces d'identification:

Numero du cours_____ Lettre de la Section_____Numero Assigné d'Étudiant_____

L'Évaluation du Dessin de l'art createur de l'Ingenieur:

Au haut de ce qui suit, chaque page a un ensemble de 2, 3, ou 4 figures de trois dimension. Utilisez une ou plus de ces figures pour créer deux dessin originaux qui accompliront le but général écrit d'en bas. Vous pouvez imaginer que ces figures sont faites de n'importe quelle matière que vous voulez, et aussi elles peuvent être de n'importe quelle taille que vous voulez pour chaque dessin. Elles peuvent être solide ou creux et peuvent être manipulés de n'importe quelle manière. Vous pouvez combiner les figures sur chaque page et dessinez des éléments additionels s'ils sont nécessaires pour votre dessin. Cependant, chaque figure ne peut être utilisée qu'une fois pour chaque dessin. Pour chaque page faites attention de faire le suivant:

1. Esquissez vos dessins.
2. Étiqueter chaque dessin (donnez une description----nommez le dessin).
3. Décrivez vos matières.
4. Identifiez des problèmes additionels que votre dessin pourrait résoudre.
5. Identifiez ceux (les personnes spécifiques) qui pourraient utiliser chaque dessin.

Le temps total pour cette évaluation est 30 minutes pour 3 pages, ou environ 10 minutes pour page. Utilisez votre temps comme vous voulez. Deux dessins doivent être créés pour chaque page. De plus, au moins une réponse doit être indiquée pour chaque boîte en bas de votre esquisse du dessin. Vous pouvez utiliser ou un stylo ou un crayon, ce que vous préférez.

Les Dessins qui produisent le son

L'esquisse	1	2
Description –nommez le dessin		
Décrivez les matières		
Les problèmes additionels résolus		
Les personnes qui pourraient utilizer votre dessin		

Les Dessins qui facilitent la communication

L'esquisse	1	2
Description –nommez le dessin		
Décrivez les matières		
Les problèmes additionels résolus		
Les personnes qui pourraient utilizer votre dessin		

Les Dessins qui peuvent voyager

L'esquisse	1	2
Description –nommez le dessin		
Décrivez les matières		
Les problèmes additionels résolus		
Les personnes qui pourraient utilizer votre dessin		

2.13.5 Chinese (Mandarin)

在開始之前，請提供以下3個標識：

課程號碼_____ 科信_____ 學生號碼_____

CEDA：創新工程的設計

在下列的是一組2個,3個或4個的三維圖形。請使用這些三維圖形中來生成兩個新的設計。這些三維圖形可以是你想要的任何材料,任意大小. 他們可以是實心或空心。你可以應用**每一頁**上的三維圖形,設計需要的元素。然而，各圖中只能用特定的三維圖形一次。

每一頁上一定要有：

1。畫出你的設計。
2。**每一**個設計標（提供簡要**說**明你的設計）。
3。**說**明你的材料。
4。描寫你的設計可以解決其他的問題。
5。描寫**每一**個設計的用戶（特定的人）。

這個測試總時間為30分鐘，共3頁，**每**頁約10分鐘。**每**頁您可以運用這個時間應建立兩種設計。此外，您可以使用鋼筆或鉛筆在下面的框中描寫您的**每一**個設計。

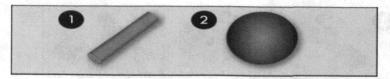

產生聲音的設計

設計畫	1	2
描述你的設計		
說明材料		
解決問題		
解釋什麼樣的人可以使用您的設計		

方便溝通的設計

設計畫	1	2
描述你的設計		
說明材料		
解決問題		
解釋什麼樣的人可以使用您的設計		

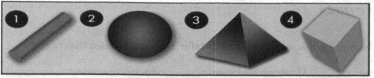

可以移動的設計

設計畫	1	2
描述你的設計		
說明材料		
解決問題		
解釋什麼樣的人可以使用您的設計		

2.13.6 German

Bevor sie beginnen geben sie bitte folgende 3 Informationen zur Identifizierung an:

Kursnummer _____ Abteilungsbuchstabe _____ Studentennummer _____

CEDA: Creative Engineering Design Assessment. Kreative Beurteilung des Entwicklungsentwurfes

Jede Seite besteht aus einem Satz von 2,3 oder 4 dreidimensionalen Figuren.

Verwenden Sie bitte eine oder mehrere dieser Figuren, um zwei **originale** Entwürfe zu erzeugen, welche Ihnen helfen das allgemeine Ziel zu erreichen, das unter diesen geschrieben steht

Sie dürfen sich die Figuren aus jeglichem Material und in jeder Größe ausdenken, welche Sie sich für jeden Entwurf vorstellen können

Sie dürfen fest oder hohl sein und können auf jede Art und Weise wie Sie möchten verstellt warden.

Sie dürfen die Figuren auf jeder Seite miteinander kombinieren und können zusätzliche Elemente zeichnen, wenn es Ihr Entwurf erfordert.

Jedoch kann jede Figur nur einmal pro Entwurf genutzt werden.

Stellen Sie sicher, dass Sie auf jeder Seite:

1. -Ihren Entwurf zeichnen
2. -Jeden Ihrer Entwürfe beschriften (geben Sie eine Kurzbeschreibung über Ihren Entwurf)
3. - Ihre Materialien beschreiben.
4. -Zusätzliche Problemstellungen beschreiben, die zur Lösung Ihres Entwurfs beitragen können.
5. - Die Benutzer (spezielle Personen) von jedem Entwurf bezeichnen.

-Die Gesamtheit dieser Bewertung beträgt 30 Minuten für 3 Seiten, oder 10 Minuten pro Seite.

-Sie können sich Ihre Zeit nach eigenem Ermessen einteilen.

-Es sollten zwei Entwürfe pro Seite erstellt werden

-Des Weiteren sollte immer eine Antwort für jeden Kasten, unterhalb Ihrer Zeichnung, pro Entwurf stehen.

-Sie können einen Füller oder Bleistift verwenden, was auch immer Sie bevorzugen.

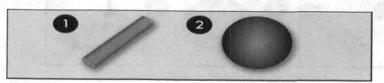

Entwürfe die Geräusche erzeugen

Entwurf/ Konzept	1	2
Beschreibung Ihres Entwurfs		
Beschreiben Sie die Materialien		
Zusätzliche Problemlösun gen		
Benutzer (Personen, welche den Entwurf nutzen können)		

Entwürfe die die Kommunikation erleichtern

Entwurf/ Konzept	1	2
Beschreibung Ihres Entwurfs		
Beschreiben Sie die Materialien		
Zusätzliche Problemlösun gen		
Benutzer (Personen, welche den Entwurf nutzen können)		

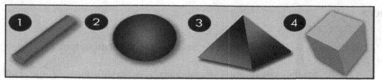

Entwürfe die sich bewegen

Entwurf/ Konzept	1	2
Beschreibung Ihres Entwurfs		
Beschreiben Sie die Materialien		
Zusätzliche Problemlösungen		
Benutzer (Personen, welche den Entwurf nutzen können)		

References

Bodner, G. M., & Guay, R. B. (1997). The Purdue visualization of rotations test. *The Chemical Educator, 2*(4), 1–17.

Branoff, T. J. (2000). Spatial visualization measurement: A modification of the Purdue SpatialVisualization Test—Visualization of Rotations. *Engineering Design Graphics Journal, 64*(2), 14–22.

Carter, C. S., Larussa, M. A., & Bodner, G. M. (1987). A study of two measures of spatial ability as predictors of success in different levels of general chemistry. *Journal of Research in Science Teaching, 24*(7), 645–657.

Charyton, C. (2005). *Creativity (scientific, artistic, general) and risk tolerance among engineering and music students.* (Doctoral Dissertation, Temple University, 2004).

Charyton, C. (2008). *Creativity (scientific, artistic, general) and risk tolerance among engineering and music students.* Saarbrücken: VDM Verlag Publishing.

Charyton, C., & Snelbecker, G. E. (2007). General, artistic and scientific creativity attributes of engineering and music students. *Creativity Research Journal, 19*, 213–225.

Charyton, C., & Merrill, J. A. (2009). Assessing general creativity and creative engineering design in first year engineering students. *Journal of Engineering Education, 98*(2), 145–156.

Charyton, C., Jagacinski, R. J., & Merrill, J. A. (2008). CEDA: A research instrument for creative engineering design assessment. *Psychology of Aesthetics, Creativity, and the Arts, 2*(3), 147–154.

Charyton, C., Jagacinski, R.J., Merrill, J.A., Clifton, W. & Dedios, S. (2011). Assessing creativity specific to engineering with the revised Creative Engineering Design Assessment (CEDA). *Journal of Engineering Education, 100*, 4, 778–799. http://www.youtube.com/watch?v=wA-PwiQ2IRQ

Gough, H. G. (1979). A creative personality scale for the adjective check list. *Journal of Personality and Social Psychology, 37*, 1398–1405.

Gough, H. G. (1992). Assessment of creative potential in psychology and the development of a creative temperament scale for the CPI. In J. C. Rosen & P. McReynolds (Eds.), *Advances in psychology assessment* (pp. 225–257). New York: Plenum.

Guay, R. B. (1978). *Factors affecting spatial test performance: Sex, handedness, birth order, and experience.* Paper presented at the Annual Meeting of the American Educational Research Association, Toronto, Ontario, Canada.

Guay, R. B. (1980). *Spatial ability measurement: A critique and an alternative.* Paper presented at the Annual Meeting of the American Educational Research Association, Boston, MA.

Kinsey, B., Towle, E., Hwang, G., O'Brien, E. J., Bauer, C. F., & Onyancha, R. M. (2007). Examining industry perspectives related to legacy data and technology toolset implementation. *Engineering Design Graphics Journal, 71*(3), 1–8.

Kovac, R. J. (1989). The validation of selected spatial ability tests via correlational assessment and analysis of user-processing strategy. *Educational Research Quarterly, 13*(2), 26–34.

Oldham, G. R., & Cummings, A. (1996). Employee creativity: Personal and contextual factors at work. *Academy of Management Journal, 39*, 607–634.

Plucker, J. A., & Renzulli, J. S. (1999). Psychometric approaches to the study of human creativity. In R. J. Sternberg (Ed.), *Handbook of creativity* (pp. 35–61). Cambridge: Cambridge University Press.

Richardson, J. S., & Morgan, R. F. (1990). *Reading to learn in the content areas.* Belmont: Wadsworth Publishing Co.

Roszkowski, M. J., Snelbecker, G. E., & Leimberg, S. R. (1989). Risk tolerance and risk aversion. In S. R. Leimberg (Ed.), *The tools and techniques of financial planning.* Cincinnati: The National Underwriter Company.

Scribner, S. A., & Anderson, M. A. (2005). Novice drafters' spatial visualization development: Influence of instructional methods and individual learning styles. *Journal of Industrial Teacher Education, 42*(2), 38–60.

Snelbecker, G. E., McConologue, T., & Feldman, J. M. (2001). *Cognitive risk tolerance survey,* Unpublished manuscript.

Snelbecker, G. E., Roszkowski, M. J., & Cutler, N. E. (1990). Investors' risk tolerance and return aspirations, and financial advisors' interpretations: A conceptual model and exploratory data. *Journal of Behavioral Economics, 19,* 377–393.

Sorby, S. A., & Baartmans, B. J. (1996). A course for the development of 3-D spatial visualization skills. *Engineering Design Graphics Journal, 60*(1), 13–20.

Yue, J., & Chen, D.M. (2001). Does CAD improve spatial visualization ability? *Proceedings of the 2001 American Society for Engineering Education Annual Conference & Exposition,* Albuquerque, NM.

Chapter 3
Future Directions: Uses

3.1 Importance to STEM

Highly creative people redefine problems, analyze ideas, persuade others, and take reasonable risks to help generate ideas (Sternberg 2001). Creativity is certainly among the most important human activity that provides conveniences and products of human inventiveness (Simonton 2000). Despite its importance to society, creativity has received relatively little attention in psychology compared to other research topics (Feist 1999; Sternberg and Lubart 1999). Although people have been engaged for centuries in creativity, only in the past few decades has this process been considered capable of analysis and improvement (Soibelman and Peña-Mora 2000). More recently, there is growing interest in and need for the utilization of creativity in the sciences.

Currently, creativity and critical thinking skills are incorporated into the core mission statement of many universities, educational programs, and college curriculum. However, few institutions utilize an empirical method of evaluating creativity. The published literature also suggests that creativity is likely domain specific (Kaufman and Baer 2005). Even in similar science areas such as engineering and chemistry—creativity may be different.

The CEDA is useful for assessing creativity in Science, Technology, Engineering, and Mathematics (STEM). Constructs such as general creativity and cognitive risk tolerance can also be assessed in STEM disciplines (Charyton and Snelbecker 2007; Charyton et al. (in press)). These dimensions may contribute toward a richer understanding of creative design specific to engineering. The CEDA has been demonstrated to be specifically related to engineering creativity and spatial skills while measuring aspects that are unique to engineering design.

"Our country's economic competitiveness and prosperity depend on innovative STEM-educated young people that work together to solve our problems effectively and creatively" (Brower et al. 2007). According to these authors, educators need to engage at least 70 % of the student population not just the top 10 % of students. Students entering STEM are a current vital need, not only for our country, but also for many countries. If students are taught that engineering can be fun through

C. Charyton, *Creative Engineering Design Assessment*,
SpringerBriefs in Applied Sciences and Technology,
DOI: 10.1007/978-1-4471-5379-5_3, © The Author(s) 2014

creative design, this could potentially engage more students to pursue engineering and may result in increased recruitment and retention in the Colleges of Engineering.

Creative design and its measurement may act as a catalyst to increase enrollment in STEM. Exposure to engineering and the CEDA could take place in universities, colleges, community colleges, institutes, academies, high schools, and junior high schools. The CEDA tool may also have the potential to measure the development of engineering design skills over time. The CEDA was developed to measure creative engineering design in adolescent students. In regard to "gifted" children, if the child is reading at the 8th grade level, then the CEDA may also be appropriate at younger chronological ages.

Creating an interest in STEM has become the new frontier (Harriger et al. 2008). Competitive degree programs require creativity and innovation (Harriger et al. 2008). STEM interests can also be heightened by broadening and establishing the relationship between creative and performing arts with broader STEM concepts (Reflections and Measures of STEM Teaching and Learning on K-12 Creative and Performing Arts Students). Harriger et al. (2008) suggests that designing rock guitars was successful for engaging high school students. The CEDA includes a problem to create designs that produce sound that could be useful in early STEM curricula.

Creativity is also a universal application of innovativeness that does not show favoritism toward race or ethnic boundary (Riffe 1985) nor gender. Underrepresented students' interests and performance are needed to foster skills that are prerequisites for STEM careers (Verma 2011). The CEDA has been administered to men and women of various racial and ethnic backgrounds and is suitable for diverse populations.

3.2 Usability in Engineering Educational Programs

Creativity is a necessity rather than an accessory for global competitiveness (Charyton et al. 2011). For sustainability and global competitiveness, the next generation of human capital needs to consist of creative and innovative college graduates (Charyton et al. (in press)). To address today's national and global challenges, graduates need skills that promote creativity, tolerance, and innovation (Adams et al. 2006). Fostering creativity should therefore be the cornerstone of engineering pedagogy (de Vere et al. 2010). As with most fundamentals, researchers (Cropley and Cropley 2000; Mahboub et al. 2004; Badran 2007) suggest that creativity and innovation should be promoted early in the education of engineers. The CEDA is written at the 8th grade level and could be used appropriately early in the education setting for potential future engineers.

Currently, a core mission statement of engineering education is to address creativity in the curriculum (Charyton et al. 2008). However, many argue that enhancing creativity in engineering design is still an evident need in many educational programs (Ferguson 1992; McGraw 2004). It is imperative to note that few institutions utilize an empirical method of evaluating creativity. Perhaps this could be due to the complexity, higher-order thinking skills, and ambiguity that challenge many to measure this potentially infinite construct. "The relationship between art and science, and their intersection in creative engineering techniques, is difficult to quantify" (McGraw 2004, p. 34). Measuring these aspects of invention could be of value to educators assessing engineering creativity (Redelinghuys 2001).

By establishing a foundation for both technical skills with creativity and innovation, engineering students can further develop and enhance their skills as they progress in the engineering profession (Badran 2007; Court 1998; Cropley and Cropley 2000). Originally, when students were chosen to participate in creativity fostering activities, Gluskinos (1971) theorized that students with higher grade point averages (GPA) would possess the highest level of creative ability. However, no significant relationship between GPA and creative ability was found (Gluskinos 1971). This suggests that creativity is different from standard measures of academic achievement.

The most commonly used processes to foster creativity within engineering education have been creativity training programs (Badran 2007; Cropley and Cropley 2000; Mahboub et al. 2004). These programs are designed to convey the meaning of creativity and creative skills to students (Cropley and Cropley 2000). Creativity training also teaches students creative thinking techniques, such as brainstorming (Badran 2007). Creativity programs utilize design project tasks which allow students opportunities to practice creative thinking using real-world engineering problems (Court 1998; Cropley and Cropley 2000; Mahboub et al. 2004).

We propose that administrations of the CEDA could be used over semesters or quarters in a two-course sequence. However, we have previously suggested more time than five weeks between multiple administrations (Charyton and Merrill 2009). My colleagues and I currently recommend at least 10 weeks or longer in between administrations. The CEDA could also be used to assess engineering education programs from beginning to end over more time ranges (from freshmen to senior or in increments each year). The CEDA could also be used when comparing introductory design skill with senior project design skills.

In our previous research, engineering students were similar in terms of creative engineering design (CEDA) and engineering creativity (PCT) (Charyton et al. 2011). Yet, there were no significant differences between Fundamentals of Engineering I and II students on these measures. We speculate that differences could potentially be found if there were two administrations of these creativity tests to the same students in the Fundamentals of Engineering I and II sequence. This may be a straightforward way to measure creative learning outcomes in series or sequential course curricula.

It is important to note that gender differences were found on the first version of the CEDA. Furthermore, in previous studies of the first version of the CEDA, female engineering and male psychology students scored higher (Charyton et al. 2008) and female engineering students scored higher (Charyton and Merrill 2009).

In more recent findings with the revised CEDA, there were no gender differences (Charyton et al. 2011).

We also found similarities in visualization spatial skills between male and female engineering students (Charyton et al. 2011), unlike past studies indicating that males tended to perform better than females (Guay 1978; Kinsey et al. 2007). We found that male and female engineering students had similar levels of visualization spatial skills that were equivalent for engineering design abilities.

We speculate that female engineering students performing as well as male engineering students may have to do with women having more access to educational resources since 1978. Historically, women were not encouraged or supported to pursue higher education. At Ohio State University, a *Women in Engineering Program* was established in 1979 to assist women engineering students. Resources are available to women engineering students (as well as men engineering students) through the *Women in Engineering Program*. More research is needed addressing spatial performance across genders. Women are still underrepresented in engineering (in many STEM disciplines including physics). Currently, women are currently only about 20 % of the engineering student body; therefore, more women still need to pursue engineering as a college major and vocation.

Further assessment of engineering design creativity is still needed, both for practical reasons and for understanding the nature of creativity specific to engineering design. Continued research of the CEDA, in conjunction with the exploration of various engineering curricula, may benefit the educational programming of universities and lead to a deeper understanding of the constructs and mechanisms necessary for creativity in engineering design. Outcomes of the CEDA may enhance our understanding of the creative processes necessary for engineering design.

The CEDA has advantages over previous measures of engineering creativity. Important components of engineering design include usefulness, functionality (Cropley and Cropley 2005; Nickerson 1999), problem finding, and problem solving (Runco 1994). The CEDA systematically assesses these essential components and cognitive processes.

The CEDA can also be used as a tool to measure the effects of the curriculum changes on student creative design. By incorporating creativity into the engineering curriculum, we may increase the frequency and quality of inventiveness. Both convergent thinking and divergent thinking are key to creativity in engineering design. The creative process incorporates these constructs. The CEDA is a useful tool that measures *convergent thinking* and *divergent thinking* which are valuable to the engineering curriculum. Creativity researchers and engineering educators may complement each other's effort toward both the awareness and measurement necessary for understanding engineering design. The CEDA offers a systematic method to assess changes and progress with learning creative engineering design in the curriculum.

Engineering education also understands the importance of using project-based learning curricula with other cultures and countries as a necessary prerequisite for completing the student's real-world preparation (Dym et al. 2005). The CEDA has potential for cross-cultural comparisons in order to understand creative engineering design across cultures. The CEDA can be an indicator of academic progress, problem solving, problem finding, and spatial skills that are necessary for creativity and innovation in many cultures. The CEDA is a tool that can be used with diverse project-based curricula and has been translated into Ukrainian, Spanish, Korean, French, Chinese, and German to address these needs.

3.3 Use in Industry

In this economic climate, the state of the industry depends on design innovation (Desposito 2009).

Universities need to collaborate with industry to enhance technological and economic competition (Burnside and Witkin 2008). Ohio State University President, Gordon Gee stated in the *Columbus Dispatch* that collaboration is more effective than competition and that universities, industries, and businesses need to collaborate. Difficulties negotiating such collaborations can be a barrier to competitiveness. Collaborations often benefit both parties involved.

Creativity and innovation are key factors for economic growth (Coconete et al. 2003). Successful organizations have learned to innovate via idea support and risk taking, which are both needed for climate building and innovation (Zeisler 2002). Successful organizations also conduct a risk analysis before implementing a creative idea (Coconete et al. 2003).

Investment in research and development are vital for creativity and innovation (Coconete et al. 2003). Creativity and innovation are crucial to engineers; therefore, researchers must ascertain the factors that motivate or create obstacles for innovation. The optimum period for stimulating creativity and innovation, either in the education or in the workforce, must also be determined (Badran 2007; Gawain 1974). For creativity to flourish, multiple collaborative perspectives are needed (Mauzy and Harriman 2003). It is also important to determine which techniques or practices will best foster creativity and innovation and how will these practices in fact benefit the future of the engineer.

As experts in their field, professional engineers are expected to provide technically accurate and efficient solutions to problems (Badran 2007; Coates 2000; Court 1998; Gawain 1974). Creativity and innovation help ensure that design solutions are original and adaptive to new engineering problems. For example, designing an automobile engine that can produce lower carbon emissions helps adapt to needs of the environment and the consumer (Court 1998).

Although some creativity skills may be taught to engineers in the educational setting, often the main focus is on technical skills such as math and physics; therefore, it is crucial to also incorporate creativity and innovation in the professional field (Badran 2007; Cropley and Cropley 2000; Mao et al. 2009; Tornkvist, 1998).

Tornkvist (1998) suggested that engineers use a method called "drill and practice," defined as repeating practice tasks or problems in order to continually come up with new creative solutions. By using practice problems and repetitive design tasks, engineers can refine their creative skills and continue to produce innovative designs. The basic interaction of engineers with their fellow employees can also influence creative thinking. For example, brainstorming sessions for design ideas combine each engineer's ability to generate more innovative solutions that are reflective of various sources of information (Salter and Gann 2003).

Mao et al. (2009) and Mich et al. (2004) suggest the use of specific methods to stimulate creativity and innovation among professional engineers. The Theory of Inventive Problem Solving (TRIZ), an acronym in Russian, includes forty inventive principles that eliminate technical conflicts between parameters of a system (Mao et al. 2009). Mao and colleagues stated that if engineers choose any of the forty inventive principles and apply that principle to solve a design problem, the engineer will provide a more innovative solution (Mao et al. 2009). For example, an engineering team was tasked with designing a wastewater treatment tunnel and encountered a conflict with the tunnel's design due to exposure of methane gas from the sewage. Using the TRIZ principle of "convert harm into benefit," the engineering team was able to design a mechanism that harnessed the methane gas as a clean energy source (Mao et al. 2009). The engineering team was able to design an innovative treatment tunnel that ultimately profited their clientele.

With the increasing dependence on technology in the modern world, researchers are also interested in using computer technology to enhance creativity and innovation in engineers. Mich and colleagues created a software program named "EPMCreate," based on the Elementary Pragmatic Model, an analytical process that helps determine interactions and relationships between two or more ideas (Mich et al. 2004). "EPMCreate" allowed engineers to input the requirements of a design project as well as the viewpoints of engineers and their clientele. This resulted in a matrix of solutions that reflected all viewpoints and requirements (Mich et al. 2004). "EPMCreate" encouraged the engineers to think about the design problem from several different perspectives to create a solution that was innovative and yet appropriate to the requirements.

Researchers such as Mao et al. (2009) and Mich et al. (2004) have developed methods that not only stimulate creativity in professional engineers, but also foster the creative thinking skills that they may have not been exposed to as a part of the traditional engineering education curriculum. For over 30 years, there has been a vital need for creativity and innovation to be a part of the professional field so that engineers can contribute their original and efficient ideas (Gawain 1974). Whether using technology, or face-to-face interactions, creativity and innovation are key for the contribution of novel and useful ideas in the engineering profession.

The engineering consultants and design firms that are most successful are those that are able to provide creative and adaptive solutions that can adjust with ever-changing times (Petre 2004; Salter and Gann 2003). As a field, engineering must be creative and innovative in order to design solutions that will meet these constantly varying problems while simultaneously being relevant and appropriate (Badran 2007; Cropley and Cropley 2000; Gawain 1974).

The USA needs a strong economy based on innovation. Ken Robinson discussed the current US crisis, "If you look at the mortality rate among companies, it's massive. America is now facing the biggest challenge it's ever faced—to maintain its position in the world economies" (Azzam 2009, p.24).

Innovation can be thought of as a practical application of creativity. However, creativity is the prerequisite for innovation. To have inventors, we need innovation. To have innovation, we need creativity. At an immediate level, creativity is essential to the engineering design process.

In design, a critical first step is to match the designer's operating mental model of the system to the end user. Human-centered design is necessary for this process. Engineering design may proceed more safely through the use of creative problem solving. Norman (2002) relates the example of a car stereo system: the "fader" control, used to adjust the balance of sound from the front and rear speakers may be indexed by turning a knob to the right or left. The design engineer, using the analog dial knob, thinks of the adjustment in a right/left axis. The user's intuitive orientation thinks of the adjustment in a horizontal front/back axis because the speakers are either in front of or in behind the driver position. The resulting product has now created more problems for user system cognition than just solving the problem of including a fader control. The designer's mental model prioritized available controller parts and spatial part orientation before the user needs. The expert design engineer must think on a different level—that of the layperson who has never used the system before (Norman 2002).

The CEDA is a valuable tool for human-centered design and can provide key information for assessing the usefulness of designs. By providing a method for assessing creativity in engineering design, educators can enable students to develop their talents as future innovative engineers.

Creative engineering exists throughout society. Virtually every man-made object or system in existence or in the past was born of an experimental, purpose-driven process to improve or create something from nothing where, if successful, the whole is greater than the sum of the parts. For example, in 2005, Motorola introduced the Razr, a cell phone dramatically thinner than all competitive units that sold 12.5 million units in less than a year. To implement this product, the development group had to move all associated teams to an innovation laboratory (Moto City) some miles away from the main headquarters in Chicago. The innovation laboratory concept helped to break barriers by focusing teams in one smaller, neutral environment (Weber et al. 2005). Hewett Packer was able to negotiate contracts with the University of California, Berkeley, and the University of California, Davis (Burnside and Witkin 2008), as collaborative partners to enhance technology.

The CEDA also may enhance design in the aviation process by measuring creativity that is specific to engineering design. Lockheed Martin fighter jets have been used since World War II, when the B-24 helped the USA end this World War.

More recently, their F-35 aircraft has been used by the United States Air Force, United States Marines, and United States Navy. The F-35 Joint Strike Fighter, designed and built by Lockheed Martin, incorporated a revolutionary design for Short Takeoff and Vertical Landing (STOVL) using a shaft-driven fan in the forward section of the aircraft compared to the competitive and existing Harrier design using redirected aft thrust. The creator, Paul Bevilaqua, had been working on the idea for 20 years (Pipinich 2006). They currently need support for men and women to access to the world's only true 5th generation multirole fighter. The newest innovations, the experiences of military forces, and the expertise of worldwide teams can help ensure that men and women in the military have the tools they need.

Innovation, Design Engineering Organization (IDEO), formed in 1991 by a merger of three design and engineering firms in Palo Alto, California, has helped to change the way product development and innovation is done (Stone 2003). Instead of designing just one product, the company specializes in developing environments, systems, and customs that may pave the way for new product lines. The human experience is often at the top of their list of priorities. Examples include the interior of the east coast corridor Acela Express high-speed train and Boston's Memorial Hospital where the experience of the nurses was prioritized at the level of the patients. Redesigning the nurse stations for more privacy and function and staging areas for patient families with improved way-finding dramatically changed the experience for the visitors and staff (Stone 2003).

Creative talent is the foundation for the development of creative industry (Yin 2009). Road mapping can improve product development (Probert and Radnor 2003). In particular, customers play a role in the creative process (Coconete et al. 2003). Understanding the market, product drivers, product attributes, product plan, technology roadmap, costs, and risks have been beneficial for systematic integration within Rockwell Automation (Probert and Radnor 2003).

The creative process requires employees to be inquisitive and gain knowledge. Open-ended questions benefit companies that strive to gain insight about the creative process (Andriopoulos 2003). Furthermore, more employees need experiences that lead toward increased creativity. Companies should also encourage their employees to be creative and innovative by rewarding their creative behavior (Shlaes 1992). High levels of creativity that achieve growth and profitability for corporations also need to be rewarded (Chu et al. 2004). For example, Hewlett-Packard's software solutions emphasize staying creative and flexible, while Net-Genesis focuses on the creative abilities of the individual in an egalitarian manner (Mauzy and Harriman 2003).

Skills such as understanding the problem, product, and users are components of the CEDA that would be directly relevant to industry. Research and development of the CEDA tool could be a useful in a variety of settings.

3.4 Use in NASA and the Military

Past studies have demonstrated usefulness for engineering design in the United States Air Force and National Aeronautics and Space Administration (NASA). Goals of the military are to have a global reach and a global presence (Cummings and Hall 2004). Both the Mars Exploration Program and Pathfinder utilized creativity for NASA's Jet Propulsions Laboratory in Pasadena, California (Shirley 2002), which employed creative teams for a hands-on robotic engineering project.

Costs are often a consideration of design (Camarda 2008). The CEDA offers the measurement of hands-on process exercises that could be useful and cost efficient. The CEDA has been translated into other languages and can be used globally. Aerospace is an important industry in the military, and the CEDA could be used at Air Force academies.

Infrastructure security such as telecommunications, finance, energy, transportation, and essential services (Bishop and Frincke 2003) is conceptualized in the CEDA problems. Security needs to take account for creative, motivated, and logical processes that may be out of the norm (Bishop and Frincke 2003). The CEDA helps the test taker think differently by generating alternative, potential solutions. Furthermore, the test taker is asked to generate additional problems solved as a component of each design.

The aeronautics and space programs of the USA, as well as other space-faring nations, strive to design products that improve services (Noor and Venneri 1998a, b). The CEDA tool also has a problem that specifically assesses designs that travel. This problem, as well as the entire CEDA, could be directly useful to NASA as well as other countries with an existing space program.

Even high-tech industries have cost constraints (Noor and Venneri 1998a). The problem NASA, industry, and academia face is that early decisions commit 90 % of costs when there is only 10 % of knowledge (Goldin 1999). These challenges are so great, not just for NASA, but for the majority of us (Goldin 1999). The CEDA tool is designed to be insightful and cost effective.

Due to the complexities and challenges of measuring creativity, there have been different approaches for assessment (Charyton 2005, 2008; Charyton et al. 2011) and evaluation (Carroll and Latulipe 2009). Carroll and Latulipe (2009) suggest that the Creativity Support Index (CSI) may be useful and was modeled after the NASA Task Load Index (TLX). The CSI contains Likert scale questions that aim to measure Czikszentmihalyi's *flow* construct that focuses on being absorbed in the creative process. The CEDA could be used along with these other measures for NASA.

NASA has recently developed conceptual design studies to facilitate geometry-centered design methodology (Fredericks et al. 2010). The CEDA contains geometry-centered basic shapes that can be manipulated in any manner (without duplication). Designs that are created could reflect aeronautics design tasks.

NASA faces challenges like industry such as cost, time, safety, and quality products (Goldin 1999). These factors are not only important for NASA and industry but also for engineering education (Goldin 1999). The CEDA incorporates these factors in the design process. Also, environments of potential spacecraft are uncertain (Goldin 1999). The CEDA measures problem finding which is generating alternate problems that could exist in order to more accurately and effectively troubleshoot. The CEDA is a paper-and-pencil assessment that also identifies potential users. When designing communication devices, the person taking the CEDA can consider possibilities such as potential users and other aspects of the design process.

Tools are needed for engineering that enhance productivity, creativity, and foster innovation from product to mission development (Noor and Venneri 1998b). Creativity is the vehicle and foundation for innovation. The CEDA is designed to measure creativity specific to engineering design. This is a skill that can be taught and can be measured. Through measurement, outcomes may include enhancing creative engineering design skills for the military and NASA. These skills may be transferable from the process that is assessed by the CEDA tool.

3.5 Conclusions

Engineering technology is key to the national economy (Lei 2010). In today's economy, creativity and competitiveness are economic drivers of innovation around the world (Florida 2005a). Innovation encompasses the development of new technologies and economic growth (Cropley 2006). Innovation is based on creativity and entrepreneurship. Creativity is the first step toward innovation (Cropley 2006). Industry's toughest decisions focus on maintaining competitive during these crucial times (Berglund et al. 2011). Design is key for enhancing innovation in our global economy.

"Today's problems can't be solved by yesterday's solutions" (Steuver 1992, p. 209). Creative engineers are driven to seek uniqueness, have unusual ideas, and tolerate unconventional thinking (de Vere et al. 2010). Cultural and creative industries are the product of the current economic times (Yin 2009). Creative economies also need to rely on fewer natural resources (Tian and Gao 2011). Based on the strong industry concern, engineers face an even greater recognition of the need to promote innovativeness. Creativity is the framework for much of innovation.

Technology, talent, and tolerance are the 3 big Ts that lead toward creativity and innovation (Florida 2005a). Wherever creativity and talent go, innovation and economic growth are sure to follow (Florida 2005b). The USA has been following with job creation and best jobs (at 11th in the world); however, has been leading with innovation and the global economy (#1 in the world) (Florida 2005b). Through prioritizing creativity and innovation, as we did with scientific creativity in the 1950s, we can lead by global prosperity, not only for the USA but for other countries also.

The word "engineer" has different meanings (Allen and Self 2008). "Ingeniatorum" from the Roman times meant someone who was "ingenious" with "gen" referring to creation, or "Genesis." The essence of the words "creativity," "create," and "engineer" stems from the act of creation.

Engineering is an applied science, a science, and an art. Creativity is a skill that all engineers can learn and practice, if they choose. These persons need hands-on opportunities to practice design skills. It is a disservice not to give engineering students and engineers an opportunity to develop and use their creativity skills. However, engineering students should not be forced to be creative. They should have the opportunity to be creative and take creativity classes if they choose. Nonetheless, both engineering and creativity are skills that can be developed over time. It is also imperative to note that persons in the workforce can incorporate engineering design skills and practice, if they choose. Employees should not be threatened to be creative or be at risk for losing their job if they are not. The CEDA is a tool that provides hands-on assessment of engineering creativity, engineering design, and most of all creative engineering design.

Oftentimes, for innovation in engineering, people of various backgrounds work in teams. Literature suggests that when working in teams, each development team needs to work within a critical mass of resources, constraints, support, and environments that encouraged creative, unconventional thinking (de Vere et al. 2010). The CEDA can also be used in team environments.

I conceptualize creativity as a source for good. Purposes for creativity as well as the CEDA should be a force for good in order to benefit humankind that is used to enhance people's lives, their health outcomes, and the prosperity of our nation as well as other countries.

Cross-cultural awareness is important for a richer understanding of creative engineering design. Furthermore, efforts should be primarily to promote sustainability as well as aviation and space travel. The CEDA has potential uses for educational programs, industry, NASA, and the military. The CEDA is a cost-effective tool through assessing creativity specific to engineering design. Through usefulness, appropriateness, constraint satisfaction, and human-centered design when considering users, the CEDA acts as a tool for assessing creative problem solving and creative problem finding.

Best wishes to you with your endeavors and prosperous uses for the CEDA. May you enhance people's lives in your projects with the CEDA.

References

Adams, R., Aldridge, D., Atman, C., Barker, L., Besterfield-Sacre, M., Bjorklund, S., et al. (2006). The research agenda for the new discipline of engineering education. *Journal of Engineering Education, 95*, 259–261.

Allen, J. E., & Self, A. (2008). Analysis of the integration of knowledge and novelty in creative engineering design. *Proceedings of the Institution of Mechanical Engineers, Part G: Journal of Aerospace Engineering, 222*(G1), 127–167. doi:10.1243/09544100JAERO237

Andriopoulos, C. (2003). Six paradoxes in managing creativity: an embracing act. *Long Range Planning, 36*(4), 375–388.

Azzam, A. (2009). Why creativity now: a conversation with Sir Ken Robinson. *Educational Leadership, 67*(1), 22–26.

Badran, I. (2007). Enhancing creativity and innovation in engineering education. *European Journal of Engineering Education, 32*(5), 573–585.

Berglund, A., Klasen, I., Hanson, M., Grimheden, M. (2011). Changing mindsets: improving creativity and innovation in engineering education. In *13th International Conference on Engineering and Product Design Education, E and PDE 2011,* Sept 8, 2011–Sept 9, 2011, (pp. 121–126).

Bishop, M., & Frincke, D. (2003). Information assurance the West Point way. *IEEE Security & Privacy,* 64–67.

Brower, T., Grimsley, R., Newberry, P. (2007). Stem is not just a four individually lettered word. In *114th Annual ASEE Conference and Exposition, 2007,* June 24, 2007–June 27, 2007, Dassault Systemes; HP; Lockheed Martin; IBM; DuPont; et al.

Burnside, B., & Witkin, L. (2008). Forging successful university-industry collaborations. *Research Technology Management, 51*(2), 26–30.

Camarda, C. J. (2008). A return to innovative engineering design, critical thinking and systems engineering. In *29th International Thermal Conductivity Conference, ITCC29 and the 17th International Thermal Expansion Symposium, ITES17,* June 24, 2007–June 27, 2007 (pp. 3–42).

Carroll, E. A., & Latulipe, C. (2009). The creativity support index. In *27th International Conference Extended Abstracts on Human Factors in Computing Systems, CHI 2009,* April 4, 2009–April 9, 2009 (pp. 4009–4014).

Charyton, C. (2005). *Creativity (scientific, artistic, general) and risk tolerance among engineering and music students.* (Doctoral Dissertation, Temple University, 2004).

Charyton, C. (2008). *Creativity (scientific, artistic, general) and risk tolerance among engineering and music students.* Saarbrücken: VDM Verlag Publishing.

Charyton, C. & Snelbecker, G.E. (2007). General, artistic and scientific creativity attributes of engineering and music students. *Creativity Research Journal, 19,* 213–225.

Charyton, C., & Merrill, J. A. (2009). Assessing general creativity and creative engineering design in first year engineering students. *Journal of Engineering Education, 98*(2), 145–156.

Charyton, C., Jagacinski, R. J., & Merrill, J. A. (2008). CEDA: a research instrument for creative engineering design assessment. *Psychology of Aesthetics, Creativity, and the Arts, 2*(3), 147–154.

Charyton, C., Jagacinski, R.J., Merrill, J.A., Clifton, W. & Dedios, S. (2011). Assessing creativity specific to engineering with the revised Creative Engineering Design Assessment (CEDA). *Journal of Engineering Education 100, 4,* 778–799. http://www.youtube.com/watch?v=wA-PwiQ2IRQ

Charyton, C., Snelbecker, G.E., Elliott, J.O. & Rahman, M.A. (in press). College Students' General Creativity as a Predictor of Cognitive Risk Tolerance. *International Journal of Creativity and Problem Solving.*

Chu, F., Kolodny, A., Maital, S. & Perlmutter, D. (2004). The innovation paradox: reconciling creativity & discipline how winning organizations combine inspiration with perspiration. *International Engineering Management Conference 2004* (pp. 949–953).

Coates, J. (2000). Innovation in the future of engineering design. *Technological Forecasting and Social Change, 64*(2), 121–132.

Coconete, D. E., Moguilnaia, N. A., Cross, R. B. M., De Souza, P. E., Sankara Narayanan, E. M. (2003). Creativity: a catalyst for technological innovation. In *IEMC'03 Proceedings, Managing Technologically Driven Organizations: 'the Human Side of Innovation and Change',* November 2, 2003–November 4, 2003 (pp. 291–295).

Court, A. W. (1998). Improving creativity in engineering design education. *European Journal of Engineering Education, 23*(2), 141.

Cropley, D. H., & Cropley, A. J. (2000). Creativity and innovation in the systems engineering process. In *Proceedings of the Tenth Annual International Symposium on Systems Engineering.*

Cropley, D., & Cropley, A. (2005). Engineering creativity: a systems concept of functional creativity. In J. C. Kaufman & J. Baer (Eds.), *Creativity across domains: faces of the muse* (pp. 169–185). Mahwah: Lawrence Erlbaum Associates.

Cropley, D. H. (2006). The role of creativity as a driver of innovation. In 2006 *IEEE International Conference on Management of Innovation and Technology, ICMIT 2006,* June 21, 2006–June 23, 2006, 2 (pp. 561–565).

Cummings, D. B., & Hall, D. W. (2004). Exploiting the creative process for innovative air vehicle design. In *42nd AIAA Aerospace Sciences Meeting and Exhibit,* Jan 5, 2004–Jan 8, 2004 (pp. 527–534).

de Vere, I., Kuys, B. & Melles, G. (2010). The Importance of Design: a comparative evaluation of problem solving in engineering education. In *International Conference on Engineering and Product Design Education* Sept 2, 2010–Sept 3, 2010 (pp. 1–6).

Desposito, J. (2009). How do we get out of this mess? Try new ideas. Electronic Design. Retrieved from http://electronicdesign.com/article/communications/how-do-we-get-out-of-this-mess-try-new-ideas21306.aspx.

Dym, C., Agogino, A., Eris, O., Frey, D., & Leifer, L. (2005). Engineering design thinking, teaching, and learning. *Journal of Engineering Education,* 103–120.

Feist, G. J. (1999). The influence of personality on artistic and scientific creativity. In R. J. Sternberg (Ed.), *Handbook of creativity* (pp. 273–296). Cambridge, UK: Cambridge University Press.

Ferguson, E. S. (1992). *Engineering and the mind's eye.* Cambridge: The MIT Press.

Florida, R. (2005a). America's looming creativity crisis. *IEEE Engineering Management Review, 33*(1), 105–113.

Florida, R. (2005b). *The flight of the creative class: The new global competition for talent.* New York: Harper Collins.

Fredericks, W. J., Antcliff, K. R., Costa, G., Deshpande, N., Moore, M. D., Miguel, E. A. S., Snyder, A. N. (2010). Aircraft conceptual design using vehicle sketch pad. In *48th AIAA Aerospace Sciences Meeting Including the New Horizons Forum and Aerospace Exposition,* January 4, 2010–January 7, 2010.

Gawain, T. H. (1974). Some reflections on education for creativity in engineering. *IEEE Transactions on Education, 17*(4), 189–192.

Gluskinos, U. M. (1971). Criteria for student engineering creativity and their relationship to college grades. *Journal of Educational Measurement, 8*(3), 189–195.

Goldin, D. S. (1999). Tools of the future. *Journal of Engineering Education, 88*(1), 31–35.

Guay, R. B. (1978). *Factors affecting spatial test performance: Sex, handedness, birth order, and experience.* Paper presented at the Annual Meeting of the American Educational Research Association, Toronto, Ontario, Canada.

Harriger, B., French, M., Aikens, M., & Shade, S. (2008). Using guitar manufacturing to recruit students into stem disciplines. *2008* In *ASEE Annual Conference and Exposition,* June 22, 2008–June 24, 2008.

Kaufman, J. C., & Baer, J. (2005). *Creativity across domains: faces of the muse.* Mahwah: Lawrence Erlbaum Associates.

Kinsey, B., Towle, E., Hwang, G., O'Brien, E. J., Bauer, C. F., & Onyancha, R. M. (2007). Examining industry perspectives related to legacy data and technology toolset implementation. *Engineering Design Graphics Journal, 71*(3), 1–8.

Lei, F. (2010). Research on education of creativity and innovation. *Applied Mechanics and Materials, 34–35,* 1742–1745.

Mahboub, K. C., Portillo, M. B., Liu, Y., & Chadranratna, S. (2004). Measuring and enhancing creativity. *European Journal of Engineering Education, 29*(3), 429–436.

Mao, X., Zhang, X., & AbouRizk, S. (2009). Enhancing value *engineering* process by incorporating inventive problem-solving techniques. *Journal of Construction Engineering & Management, 135*(5), 416–425.

Mauzy, J., & Harriman, R. A. (2003). Three climates for creativity. *Research Technology Management, 46*(3), 27–30.

McGraw, D. (2004). Expanding the mind: Creativity is such an integral part of being an engineer, but how on earth do you teach it? *ASEE Prism, 13*, 31–36.

Mich, L., Anesi, C., Berry, D. M. (2004). Requirements engineering and creativity: an innovative approach based on a model of the pragmatics of communication. In *Proceedings of Requirements Engineering: Foundation of Software Quality REFSQ, 4.*

Nickerson, R. S. (1999). Enhancing creativity. In R. J. Sternberg (Ed.), *Handbook of creativity* (pp. 392–430). Cambridge: Cambridge University Press.

Noor, A. K., & Venneri, S. L. (1998a). ISE—intelligent synthesis environment for future aerospace systems. *Part 1 (of 5),* March 21, 1998–March 28, 1998, 2 (pp. 467–476).

Noor, A. K., & Venneri, S. L. (1998b). ISE provides a new frontier for synthesis of complex engineering products and missions. *Part 3 (of 5),* Oct 11, 1998–Oct 14, 1998, 3 (pp. 2698–2703).

Norman, D. A. (2002). *The design of everyday things [The psychology of everyday things].* New York: Basic Books.

Petre, M. (2004). How expert engineering teams use disciplines of innovation. *Design Studies, 25*(5), 477–493.

Pipinich, R. E. (2006). High-stakes creativity. *Industrial Engineer: IE, 38*(6), 30–35.

Probert, D., & Radnor, M. (2003). Frontier experiences from industry-academia consortia: corporate roadmappers create value with product and technology roadmaps. *Research Technology Management, 46*(2), 27–30.

Redelinghuys, C. (2001). Proposed measures for invention gain an engineering design. *Journal of Engineering Design, 11*, 245–263.

Riffe, W. J. (1985). Creativity: catalyst for education. *Motor Vehicle Technology: Mobility for Prosperity, 589–593.*

Runco, M.A. (1994). *Problem finding, problem solving, and creativity.* Westport, CT: Ablex Publishing.

Salter, A., & Gann, D. (2003). Sources of ideas for innovation in engineering design. *Research Policy, 32*(8), 1309–1324.

Shirley, D. L. (2002). Managing creativity: A creative engineering education approach. *ASEE Annual Conference and Exposition: Vive L'Ingenieur,* June 16, 2002–June 19, 2002, 2367–2380.

Shlaes, C. (1992). Rewarding and stimulating creativity and innovation in technology companies. In *Proceedings of the 1991 Portland International Conference on Management of Engineering and Technology—PICMET '91,* Oct 27, 1991–Oct 31, 1991 (pp. 609–612).

Simonton, D. K. (2000). Creativity: cognitive, personal, developmental and social aspects. *American Psychologist, 55*(1), 151–158.

Soibelman, L. & Peña-Mora, F. (2000). Distributed multi-reasoning mechanism to support conceptual structural design. *Journal of Structural Engineering, 733–742.*

Sternberg, R. J. (2001). What is the common thread of creativity? Its dialectical relation to intelligence and wisdom. *American Psychologist, 56*(4), 360–362.

Sternberg, R. J., & Lubart, T. I. (1999). The concepts of creativity: Prospects and paradigms. In R. J. Sternberg (Ed.), *Handbook of creativity* (pp. 3–15). Cambridge: Cambridge University Press.

Steuver, J. K. (1992). Creativity: a response to change and competition. In *1992 International SAVE Proceedings,* May 31, 1992–June 3, 1992, 27 (pp. 209–214).

Stone, B. (2003). Reinventing everyday life. *Newsweek, 142*(17), 90–92.

Tornkvist, S. (1998). Creativity: Can it be taught? The case of engineering. *European Journal of Engineering Education, 23*(1), 5–12.

Tian, Y., & Gao, C. (2011). Management strategies of creative industries uncertainty. In *2011 International Conference on Business Computing and Global Informatization*, July 29, 2011–July 31, 2011 (pp. 79–82).

Verma, A. K. (2011). Attracting K-12 students towards engineering disciplines with project based learning modules. In *118th ASEE Annual Conference and Exposition*, June 26, 2011–June 29, 2011.

Weber, J., Holmes, S., & Palmeri, C. (2005). 'Mosh pits' of creativity. *Business Week, 3958*, 98–100.

Yin, Q. (2009). Talking about design and cultural and creative industry. In *2009 IEEE 10th International Conference on Computer-Aided Industrial Design and Conceptual Design: E-Business, Creative Design, Manufacturing—CAID and CD'2009*, Nov 26, 2009–Nov 29, 2009 (pp. 2159–2162).

Zeisler, S. (2002). From quality to break through: New heights in innovation. *56th Annual Quality Congress Proceedings*, May 20, 2002–May 22, 2002, 543–547.

About the Author

Christine Charyton, PhD, is a visiting assistant professor in the Department of Neurology at the Ohio State University Wexner Medical Center and a lecturer in the Department of Psychology at the Ohio State University in Columbus, Ohio.

Dr. Charyton's research encompasses the synthesis and unification of interdisciplinary knowledge and transdisciplinary science. Dr. Charyton focuses on the integration of psychology and engineering with cognitive and learning interventions. Dr. Charyton has been actively working on the measurement and assessment of creative engineering design and cognitive risk tolerance.

Dr. Charyton is also a licensed psychologist in private practice specializing in the treatment of comorbid psychological and neurological conditions using cognitive behavioral therapy (CBT) in combination with creative interventions such as mindfulness and positive psychology. Dr. Charyton's clinical research integrates interventional research with neuroscience and epidemiology.

C. Charyton, *Creative Engineering Design Assessment*,
SpringerBriefs in Applied Sciences and Technology
DOI: 10.1007/978-1-4471-5379-5, © The Author(s) 2014